Microbial Removal and Integrity Monitoring of High-Pressure Membranes

The mission of the Awwa Research Foundation (AwwaRF) is to advance the science of water to improve the quality of life. Funded primarily through annual subscription payments from over 1,000 utilities, consulting firms, and manufacturers in North America and abroad, AwwaRF sponsors research on all aspects of drinking water, including supply and resources, treatment, monitoring and analysis, distribution, management, and health effects.

From its headquarters in Denver, Colorado, the AwwaRF staff directs and supports the efforts of over 500 volunteers, who are the heart of the research program. These volunteers, serving on various boards and committees, use their expertise to select and monitor research studies to benefit the entire drinking water community.

Research findings are disseminated through a number of technology transfer activities, including research reports, conferences, videotape summaries, and periodicals.

Microbial Removal and Integrity Monitoring of High-Pressure Membranes

Prepared by:
Jim Lozier, **Mehmet Kitis**, and **Christian Colvin**
CH2M HILL Inc.
Englewood, CO 80112

and

Jae-Hong Kim, **Baoxia Mi**, and **Benito Mariñas**
University of Illinois at Urbana-Champaign
Department of Civil and Environmental Engineering
Urbana, IL 61801

Jointly sponsored by:
Awwa Research Foundation
6666 West Quincy Avenue
Denver, CO 80235

and

U.S. Department of the Interior
Bureau of Reclamation
1849 C Street NW
Washington, DC 20240

Published by the

Awwa Research Foundation
Advancing the Science of Water®

IWA Publishing

DISCLAIMER

This study was funded by the Awwa Research Foundation (AwwaRF). AwwaRF assumes no responsibility for the content of the research study reported in this publication or for the opinions or statements of fact expressed in the report. The mention of trade names for commercial products does not represent or imply the approval or endorsement of AwwaRF. This report is presented solely for informational purposes.

Published by IWA Publishing, Alliance House, 12 Caxton Street, London SW1H 0QS, UK

Telephone: +44 (0) 20 7654 5500; Fax: +44 (0) 20 7654 5555; Email: publications@iwap.co.uk;
Web: **www.iwapublishing.com**

AwwaRF report number 90942F. Originally published by AwwaRF for its subscribers in 2003.
IWA Publishing version published 2004.

ISBN 1 84339 850 8

CONTENTS

TABLES

FIGURES

FOREWORD

The Awwa Research Foundation is a nonprofit corporation that is dedicated to the implementation of a research effort to help utilities respond to regulatory requirements and traditional high-priority concerns of the industry. The research agenda is developed through a process of consultation with subscribers and drinking water professionals. Under the umbrella of a Strategic Research Plan, the Research Advisory Council prioritizes the suggested projects based upon current and future needs, applicability, and past work; the recommendations are forwarded to the Board of Trustees for final selection. The Foundation also sponsors research projects through the unsolicited proposal process; the Collaborative Research, Research Applications, and Tailored Collaboration programs; and various joint research efforts with organizations such as the U.S. Environmental Protection Agency, the U.S. Bureau of Reclamation, and the Association of California Water Agencies.

This publication is a result of one of these sponsored studies, and it is hoped that its findings will be applied in communities throughout the world. The following report serves not only as a means of communicating the results of the water industry's centralized research program but also as a tool to enlist the further support of the nonmember utilities and individuals.

Projects are managed closely from their inception to the final report by the foundation's staff and large cadre of volunteers who willingly contribute their time and expertise. The foundation serves a planning and management function and awards contracts to other institutions such as water utilities, universities, and engineering firms. The funding for this research effort comes primarily from the Subscription Program, through which water utilities subscribe to the research program and make an annual payment proportionate to the volume of water they deliver and consultants and manufacturers subscribe based on their annual billings. The program offers a cost-effective and fair method for funding research in the public interest.

A broad spectrum of water supply issues is addressed by the foundation's research agenda: resources, treatment and operations, distribution and storage, water quality and analysis, toxicology, economics, and management. The ultimate purpose of the coordinated effort is to assist water suppliers to provide the highest possible quality of water economically and reliably. The true benefits are realized when the results are implemented at the utility level. The foundation's trustees are pleased to offer this publication as a contribution toward that end.

Edmund G. Archuleta, P.E.
Chair, Board of Trustees
Awwa Research Foundation

James F. Manwaring, P.E.
Executive Director
Awwa Research Foundation

ACKNOWLEDGMENTS

This project was funded jointly by the Awwa Research Foundation and the U.S. Bureau of Reclamation.

The authors of this report are grateful to the Orange County Water District (OCWD), Fountain Valley, CA and the City of Chesapeake (COC), VA for their cooperation and for providing the project team use of facilities at the Water Factory 21 and Northwest River Water Treatment Plant during the conduct of the pilot-scale tests. In particular, the assistance of Mehul Patel and Betsy Martin of OCWD and Karen Harr of COC are greatly appreciated. The authors would also like to recognize the efforts of Kristi Perri and Carrie Adam for conducting the pilot testing at the Northwest River Water Treatment Plant and Dave Munzer for his invaluable assistance in making the pilot equipment at this location operational. All are with CH2M HILL's Hampton Roads, VA office.

The authors would like to thank Courtney Acker (with Carollo Engineers) who conducted some portion of the bench-scale experiments at the University of Illinois, Urbana-Champaign during her master's research.

EXECUTIVE SUMMARY

INTRODUCTION

As early as 1972, membrane processes were viewed as a potential barrier to microorganisms in water treatment. Not until the promulgation of the Surface Water Treatment Rule (SWTR) in 1989, did the United States Environmental Protection Agency (EPA) establish treatment techniques for controlling viruses, bacteria, and the protozoa *Giardia lamblia* in drinking water supplies. In its Guidance Manual for the SWTR, EPA classified membrane filtration processes as alternative technologies, and required that pilot studies or other means be used to demonstrate that these technologies meet the requirements of the SWTR. The exception was reverse osmosis (RO), which was considered by EPA to be effective for removing *Giardia* cysts and viruses. Consequently, demonstration was considered unnecessary. This consideration was based primarily on the nominal cutoff of RO membranes (0.001 microns [μm]) relative to the size of the pathogens of concern (0.025 μm to 10 μm)[1]. However, at the time of SWTR promulgation, all RO facilities located in the U.S. were treating brackish groundwater (and in some instances, surface water) supplies for the purpose of dissolved solids reduction (demineralization). Pathogen removal was not a treatment requirement for these plants and no disinfection credit was requested by utilities or granted by state regulators. With the advent of the interim and long-term Enhanced Surface Water Treatment Rules (ESWTR), an increased emphasis has been placed on the control of persistent pathogens in drinking water supplies, particularly those that are resistant to inactivation by traditional disinfection methods, specifically *Giardia lamblia* cysts and *Crytosporidium parvum* oocysts. As RO and NF are used increasingly to treat brackish and organic laden surface waters, the need to obtain disinfection credits becomes more critical. Further, the expanding application of RO for advanced reclamation of municipal wastewater effluent for the purpose of indirect potable reuse has placed increased emphasis on the removal capability of RO for pathogens. In California, indirect potable reuse via surface water supply augmentation requires that RO provide 4-log removal of virus (State of California, Dept. of Health Services).

Beginning in 1993, low-pressure membrane filtration, microfiltration (MF), and ultrafiltration (UF) have been applied at large-scale (>1 mgd) at an increasing rate to meet the requirements of the SWTR. Microfiltration and UF have found ready acceptance by state regulatory agencies for this purpose based on numerous studies showing their ability to provide high levels (>5 logs) of *Giardia* and *Cryptosporidium* removal, and because such removal can be routinely demonstrated using direct integrity methods. Integrity of MF/UF systems for *Giardia* and *Cryptosporidium* removal can be routinely and automatically demonstrated using pressure/vacuum hold (P/VH) and diffusive air flow (DAF) tests, derived from the principle of bubble point testing. To date, P/VH and DAF tests have not been applied to high-pressure membrane filtration processes (NF and RO), except as a quality control assessment method on

[1] It has not been conclusively demonstrated that high-pressure (RO and nanofiltration [NF]) membranes have distinct pores. Many membrane theorists consider RO and NF membranes to be non-porous and the passage of solutes (e.g., sodium and chloride) to occur by the solute first dissolving in the membrane on the feed side, and then diffusing through the membrane to the permeate side, hence, the solution/diffusion theory of solute transport. For this reason, the term nominal pore size is used in a non-rigorous manner in this report and sizes of 0.001 and 0.002 μm are used somewhat loosely to characterize the nominal pore sizes of RO and NF, respectively.

individual modules (ASTM Standards, 1988). Consequently, there are no integrity methods currently in use with full-scale RO and NF systems to demonstrate pathogen removal beyond the indirect application of conductivity monitoring, which limits demonstrated removals to <2 logs. Given the capability of RO and NF to provide much greater levels of virus and *Giardia*/*Cryptosporidium* removal, it is critical that research be conducted to demonstrate such ability, assess the potential for and impact of integrity compromization in both modules and systems, and identify integrity test method(s) that could be cost-effectively applied to demonstrate integrity of full-scale systems, both initially and on an on-going basis.

RESEARCH OBJECTIVES

The main objective of this project was to determine, at both bench- and pilot-scale, the microbial integrity of high-pressure RO and NF membranes using both microbial and non-microbial surrogate indicators. This research focused on virus, rather than protozoan, removal, given that virus particles are significantly smaller and would represent a greater integrity challenge. The selected viral indicator was MS2 *phage* due to its similar size, shape, and nucleic acid composition to enteric viruses. The non-microbial surrogate indicators chosen for evaluation were 0.02-µm orange fluorescent carboxylate-modified microspheres and Rhodamine WT (R-WT), a soluble fluorescent dye having a molecular weight of approximately 480. Microspheres were selected as they are particles having a similar size to MS2 *phage*, and can be measured at low concentrations (50 nanograms per liter [ng/L]), allowing challenge levels of >4 logs using relatively low feed concentrations (~1 mg/L). Rhodamine WT was chosen because it is commonly used in water treatment plants as a tracer, is approved by the EPA for such application, and should be well rejected by RO and NF (molecular weight greater than molecular weight cutoff of both membranes). Based on recent advancements in fluorometer design, R-WT can be quantified at levels as low as 10 to 20 ng/L in clean water (permeate). This would allow a challenge level of 4 to 5 logs at a feed concentration of 0.1 to 1 mg/L, levels that are approved for use by EPA with water treatment facilities. Minimum challenge level of 4 logs for the non-microbial surrogates was selected to capture the 4-log removal requirement for viruses in the SWTR and ESWTR.

The specific objectives of this research were to:

- Determine the removal of virus, microsphere, and R-WT by intact RO and NF membranes in bench-scale (flatsheet coupons and single, small capacity spiral-wound elements) and pilot-scale test units;
- Determine the impact of different level of RO and NF membrane compromizations (through induced imperfections on the active and support layers) on virus, microsphere, and R-WT removals in bench- and pilot-scale test units;
- Determine the impact of RO and NF membrane fouling and further chemical cleaning on virus, microsphere and R-WT passage through compromised RO and NF membranes;
- Determine the level of o-ring compromization to result in the increased passage of virus, microsphere, and R-WT;
- Determine the relationship between the rejections of the microbial- and non-microbial virus surrogate indicators with that of the standard indicator of RO and NF membrane integrity, conductivity;

- Compare and evaluate the practical use of the three surrogates in full-scale integrity tests;
- Determine the impact on virus, microsphere, and R-WT removal of a compromised RO or NF element or o-ring when installed at different locations within a multiple-array (stage) membrane system;
- Assess, by visual means, the reactivity of RO and NF elements toward the uptake of tested surrogates.

APPROACH

For the bench-scale portion of this project, five membranes were investigated in flat-leaf coupon configuration, and three among them were also tested with spiral-wound elements. Three major approaches were tested for assessing membrane integrity: (1) simultaneous testing of high-rejected and low-rejected solutes, (2) use of 0.024-μm fluorescent-dyed polystyrene microspheres as non-biological surrogate indicators for fecal viruses, and (3) use of MS2 *phage* as biological surrogatc indicator for fecal viruses. Experiments were performed with membranes as received from the manufacturers as well as membranes after attempting to compromise their integrity by chemical and physical means.

The second phase of this project included pilot-scale testing conducted at the following sites:

- Orange County Water District (OCWD) Water Factory 21 (WF-21) in Fountain Valley, CA
- The Northwest River Water Treatment Plant (NWR-WTP) in the City of Chesapeake, VA

The feedwaters to the test elements in these two sites were as following:

- Synthetic water (i.e., permeate from full-scale surface water RO facility at NWR-WTP)
- Microfiltered secondary effluent (MFSE) at WF-21
- Conventionally-treated river water (CTRW) polished using 10-μm cartridge filters at the NWR-WTP

Waters from WF-21 and NWR-WTP were selected to reflect the two generic types of source waters where disinfection (removal) credit for pathogens by RO and NF would be most likely required: surface waters and reclaimed wastewater effluent.

Two types of 4-inch diameter by 40-inch length spiral-wound elements, Hydranautics Model ESPA1-4040-UHT RO and Koch Fluid Systems Model TFCS-4040S NF, were tested in the pilot work.

CONCLUSIONS

The primary conclusions drawn from this project are summarized below:

1. Intact ESPA1 RO elements (as received from the suppliers) tested both in flat-leaf and spiral-wound configurations, generally demonstrated from >6-logs to complete removal of MS2 *phage* and about 4-log removal of both R-WT and microspheres (the maximum removal capable at the concentrations applied). Under similar operating conditions, intact TFCS NF elements showed lower surrogate removal: 2.5 to 4 logs for R-WT and microspheres and between 3 and 5.5 logs for MS2 *phage*. This result was not anticipated for the NF membrane for the two particulate surrogates (MS2 *phage* and microspheres), whose rejection should be complete or near complete by the non-porous RO and NF membranes. Lesser rejection of R-WT could be expected based on the larger MWCO of the NF membrane. However, the NaCl and/or conductivity/TDS rejections of the intact NF elements were found to be within or greater than the manufacturer's specifications, indicating that conductivity/TDS rejection is not a good predictor or virus removal.
2. The largest range in log removals with the NF elements was for MS2 *phage*. One element passed approximately 4 logs of MS2 *phage*, a significant degree of passage. In contrast, the variability in surrogate removals by ESPA1 was significantly less.
3. Of the various chemical and physical approaches used to compromise membrane integrity (limited exposure to free chlorine, air entrainment, scratch of the active layer, and pinholes of various sizes) only pinholes fully penetrating both the active (rejecting) and support layer of a membrane leaf produced significant increases in the passage of both biological and non-biological surrogate indicators as well as conductivity and TDS.
4. The passages of enteric viruses through the induced membrane imperfections (i.e., pinhole) may be satisfactorily predicted by both the particulate and dissolved surrogates used in this study, as long as the size of a pinhole is sufficiently large not to cause a steric hindrance to passage of viral size particles.
5. A crack induced along the inner surface of the o-ring was insufficient to cause a loss of integrity. However, removing small sections from o-rings significantly decreased the log removals; removal decreased proportionately with increasing size of removed section.
6. For multi-stage test unit (MSTU) testing, log removal of all three surrogates, as well as conductivity and TDS, as measured in the combined permeate, decreased as compromised o-ring location was moved from front to tail end of the two-stage array. The combination of two factors, decreasing net driving force (NDF) pressures (resulting in less o-ring deformation and higher flow across the cut) and increasing feed concentrations, could explain this result. Larger pressures provided more compression on the o-ring, closing the gaps in the removed section of the o-ring thereby decreasing flow from feed to permeate. On the other hand, the impact of location of element with induced imperfection on surrogate removals varied depending on the type of surrogate. Log removals for the particulate surrogates (*phage* and spheres) increased as the compromised element was moved from vessel 1 to vessel 6. For soluble R-WT, the impact of vessel location was negligible. For the particle-based surrogates (removed by "physical sieving" mechanism) the rate of passage of the particles through the membrane imperfection was controlled more by NDF pressure than by feed concentration. In addition, the higher concentration of surrogate particles in the feed as the compromised element is moved from front to

trailing position could more effectively occlude the imperfection, thereby reducing the rate of advective flow through the imperfection. Soluble R-WT, removed by "solution-diffusion" mechanism, would not be expected to cause occlusion and reduce advective flow.

7. Fouling, presumably through the bridging of holes in the membrane layer by feedwater particles, can "heal" minor imperfections present from manufacture as well as the larger imperfections induced in this research, however the degree of healing will be proportional to both the membrane's propensity for fouling and the foulant characteristics of the feedwater.
8. Increases in surrogate log removal by fouling of compromised membrane were reversed by chemical cleaning. The degree of reversibility may be dependent upon the physicochemical nature of the foulant layer and the surface characteristics of the membrane active layer.
9. Virus challenge concentrations of >7 logs could be readily achieved with the MS2 *phage* culturing techniques developed in this project. Based on the high cost of the microspheres used in this study, challenge level of 4-4.5 log could be achieved using a practical feed concentration of 1 mg/L. Assuming development of lower-cost microspheres specifically designed for this application, higher challenge levels would allow for the measurement of greater log removals, thereby making microspheres a more ideal non-biologic viral surrogate.
10. Conductivity and/or TDS rejection cannot be used as an accurate predictor of viral passage. Intrinsic imperfections are not distinguishable using normal manufacturer's quality assurance/quality control (QA/QC) "wet testing", which simply measures percent rejection of a salt solution.
11. R-WT may be a practical surrogate for detecting imperfections in both RO and NF membranes relative to virus removal, although it does have limitations. R-WT is relatively inexpensive, approved for potable water use, easy to dose, and can be accurately measured at concentrations down to 20 ng/L in membrane permeates, resulting in an ability to measure about 4 log removal at a feed concentration of <2 mg/L. Because R-WT is soluble, it has a tendency to adsorb into the membrane (to a greater extent with the NF membrane evaluated here), and because of its moderate molecular weight, is not completely rejected by high pressure membranes. Maximum removal is most likely limited to 4-logs for RO and 3-logs for NF. However, it can serve as a readily available alternative to other methods currently employed that have lesser sensitivity, including conductivity and total organic carbon (TOC). Its removal is by solution-diffusion and dependent on feed concentration, which should be accounted for when comparing results between single-element test units and multi-stage systems as in full-scale applications. The estimated cost of R-WT for a single challenge test of a 5-mgd membrane facility operating at 80% recovery is less than $500.
12. Commercially available fluorescent microspheres are currently too expensive for full-scale membrane integrity testing. The estimated cost for their use in one challenge test of a 5-mgd membrane plant operating at 80% recovery is nearly $100,000.
13. The findings associated with the membrane and o-ring compromization highlight the importance of conducting integrity checks of complete high pressure membrane systems, not simply membrane elements. The impact of compromised o-ring/element

position further indicates that integrity failures are a function of not only a specific component, but also the location of that component in the system.

FUTURE RESEARCH

Research should be devoted to identifying a more suitable non-biologic, particulate viral surrogate for cost-effective, routine use with full-scale RO/NF systems. Such a surrogate is most ideal for the application compared to non-biologic, soluble compounds because it best reflects the characteristics of the virus particle. The fluorescent microspheres evaluated here are currently too expensive for full-scale use and require sophisticated instrumentation for analysis. The high cost and analytical requirements reflect manufacturing methods that must attain very high precision and QA/QC requirements necessary to satisfy their targeted end use – biomedical applications. It is very possible that through a collaborative research partnership between AwwaRF (possibly working with other research agencies), utilities and existing microsphere suppliers (or other companies having the capability to produce "tagged" particles), such a particle can be engineered that possesses the necessary attributes at significantly lower cost. Target characteristics of a "designer" microsphere would include cost effectiveness for dosing to large flows (up to 50 mgd), ability to measure at low concentrations using a readily calibrated, field instrument; and health/environment characteristics that permit approval for use with potable water (NSF 60 certification) and for discharge to surface water bodies (NPDES permit). One manufacturer has already indicated interest in the potential development of such a surrogate (Moll, 2002).

Given that development of an appropriate particulate, non-biologic surrogate may require several years, further research is needed in the short-term to confirm the applicability of Rhodamine WT (R-WT) dye as a suitable indicator of NF/RO membrane viral integrity at full scale. Such research should address the following issues: (1) ability to generalize results to the complete range of commercially available NF/RO; (2) what loss of R-WT removal occurs with aged intact membranes, specifically those that have suffered an appreciable loss of salt rejection; (3) how effective R-WT is as a tool to locate a single compromised membrane in a plant having many pressure vessels using the common vessel "probe" technique; and (4) what constraints will regulators place on its use (maximum feed dose and duration, maximum permeate concentration for distribution) and what use frequency is required to obtain viral removal credits.

Additional research should be considered to identify a viral surrogate that possesses the same advantages as R-WT but is more highly rejected by high pressure membranes, specifically NF membranes and is less adsorbable in the membrane. Candidates would be compounds of higher molecular weight (>1000 daltons) and/or more highly charged, such that 5-log removal by intact NF and RO membrane could be demonstrated. The use of computer programs that can "design" molecules based on the desired properties would be advantageous in this respect.

Given the range of virus removal capability demonstrated by the NF membrane evaluated in this research as well as among membrane types (Adham *et al.*, 1998a,b), additional research is needed to identify the primary causes of this variability. Ideally this research would be conducted jointly with the membrane suppliers. Given that all membrane types evaluated to date are formed by the same interfacial polymerization method using polysulfone support and polyamide barrier polymers, the degree of variability is surprising and needs to be understood and addressed. Fundamental understanding will do much to remove the causes of such variability or conversely identify "on-line" healing techniques. In either case, such research is necessary to further regulatory acceptance of high pressure membranes for virus removal.

APPLICATION TO WATER INDUSTRY

The applicability of this research to the water industry can be summarized as follows:

1. ***Rhodamine WT.*** Rhodamine WT (R-WT) is currently applicable as a surrogate for evaluating the integrity of full-scale high pressure membrane systems for virus removal. In this regard, R-WT has been included in the draft American Society for Testing and Materials (ASTM) draft *Standard Practice for Integrity Testing of Water Filtration Membrane Systems* (ASTM 2002). It must be noted that the capability of R-WT to assess viral removal will differ for RO than NF and could differ between manufacturer's products for each process. It is reasonable to assume that 4-log virus removal can be successfully demonstrated with a RO system, but only 3-log removal with a NF system. The reader is referred to the appropriate section of this draft practice for guidelines on the use of R-WT. The actual application of R-WT at full-scale will have to address a number of issues that were identified but not resolved during this testing. These include:
    - Approval by relevant State and Federal regulatory agencies regarding R-WT application to a full-scale membrane facility. To date, R-WT has only been approved for use for tracer studies in water treatment plants.
    - Time required to flush the system following R-WT application in order to reduce concentrations of R-WT in the membrane permeate to less than 100 ng/L (or other level imposed by regulatory agency). The flushing duration will be a function of log removal by the full-scale membrane system, most notably whether the system employs RO or NF membranes. It is anticipated that flushing time with RO membranes may be very short given that log removals by intact RO membranes as observed in this study produced permeate R-WT well below 100 ng/L. For systems with NF membranes, flushing time will have to be determined based on the NF product used and its actual R-WT rejection.
    - Ability to discharge membrane concentrate containing R-WT at concentrations up to 7 mg/L. Assuming a target R-WT dose in the membrane feedwater of 1 mg/L and a system capability of 3-4 logs R-WT removal, the concentrate from the full-scale plant will contain R-WT concentrations of 6-7 mg/L. At this concentration, the dye color may be observable. Depending upon the source of concentrate discharge, the presence of R-WT at these levels may represent a challenge, particularly for surface water discharges. The applicable regulatory agency(ies) should be contacted to determine the implications of R-WT use. It must be kept in mind that the time period for R-WT dosing can be very short (<15 minutes) and as such the mass of microspheres to be discharged will not be significant.
2. ***Fluorescent Microspheres.*** Currently, fluorescent microspheres of the type used in this research are not practical for use by utilities in assessing the integrity of full-scale membrane systems. Although the microspheres showed the best correlation to MS-2

phage with respect to both intact and compromised membranes and systems, their high cost makes their dosing of full-scale facilities prohibitive. Furthermore, current analytical methods needed for their detection require expensive and sophisticated bench-top instrumentation and well-trained technicians. These constraints reflect the development of such microspheres for the biomedical industry, where a high level of precision and quality control/quality assurance in manufacturing is needed. In order to make microspheres applicable to membrane integrity testing at full-scale, it will be necessary to engineer a product that has much lower production costs and is designed specifically for such a use. This will require collaboration between the water industry and either current microsphere manufacturers or other commercial entity with the capability and interest to produce such a product.

3. ***Types of Compromization.*** The types of compromization used in this research are not necessarily those that most utilities would be expected to encounter with full-scale plants. The quality control practiced by most membrane suppliers would make "pin pricks" rare, however, it is certainly possible that membrane imperfections caused by faulty barrier layer coating and polymerization or glue line discontinuities can produce similar advective flow. For o-rings, more common modes of failure relate to partial cuts and cracks from improper installation and shimming (where o-rings can be rolled from the groove). In this case, the results obtained in this study for the cracked o-rings are most pertinent and indicate that major loss of compromization from o-ring failures should be low.

4. ***Membrane Fouling.*** A major and important finding from this research that has direct applicability to the industry is the ability of membrane fouling to "heal" membrane imperfections and further that such healing could be reversed by chemical cleaning. It was surprising to find that highly pretreated membrane feedwater (feed filtered by a 0.2 um microfilter), where only submicron particles should be present, was capable of bridging a defect on the order of 300-500 microns. Given that most membrane imperfections induced during the membrane/element manufacturing process should be less than this size, it is reasonable to assume that viral passage through imperfections in elements received from the manufacturer should experience decreased viral passage with increased operation. This has been observed for many full-scale plants with respect to conductivity/TDS passage and is termed "membrane conditioning" in the industry. Such operational improvement in virus removal is advantageous and with sufficient sampling and statistical analysis, could provide the basis for ensuring that full-scale systems receive appropriate virus removal credit following some initial conditioning period. This, of course, assumes that the full-scale system contains sufficient imperfections as not to meet regulatory requirements upon start-up.

CHAPTER 1
A REVIEW OF INTEGRITY MONITORING OF HIGH-PRESSURE MEMBRANE PROCESSES FOR WATER TREATMENT

INTRODUCTION

This section reviews previous studies of microbial removal by high-pressure membranes, and identifies mechanisms by which viruses pass nanofiltration (NF) and reverse osmosis (RO) membrane systems. It also identifies methods to monitor membrane integrity, including methods that employ virus surrogates and means of quantifying passage of surrogates through membranes. Most methods previously used or evaluated have some limitations. Other methods examined in this report are new to water treatment, having been developed for other applications. Criteria for evaluating appropriate test methods are presented.

RESEARCH NEED

Pressure-driven membrane processes used in drinking water treatment include microfiltration (MF), ultrafiltration (UF), NF, and RO or hyperfiltration. Microfiltration and UF operate at low feed pressures (5 to 50 pounds per square inch (psi) [0.34 to 1.03 bar]). Nanofiltration and RO operate at high feed pressures (75 to 1,200 psi [5.2 to 82.7 bar]) and are the focus of this study. Other AwwaRF studies have evaluated issues associated with microbial removal by low-pressure membrane systems (Sorber, Malina, and Sagik 1972; USEPA Final Rule 1989).

As early as 1972, RO membranes were studied for their ability to remove microorganisms from water supplies (Sorber, Malina, and Sagik 1972). Not until the promulgation of the Surface Water Treatment Rule (SWTR) in 1989, did the U.S. Environmental Protection Agency (EPA) establish treatment techniques for controlling viruses, bacteria, and the protozoa Giardia lamblia in drinking water supplies (USEPA Final Rule 1989). With the passage of the SWTR, the need for control of pathogenic microorganisms in water treatment became a regulatory requirement. Although high-pressure membrane processes have seen only limited use in the treatment of surface waters, most recently increased interest in their application has developed with the promulgation of the Disinfectants/Disinfection Byproduct Rule (D/DBP) and the interim Enhanced Surface Water Treatment Rule (ESWTR). The ability of high-pressure membranes to effectively remove both pathogens and disinfection byproduct precursors make them ideal processes for addressing these conflicting regulations.

In its Guidance Manual for the Surface Water Treatment Rule (USEPA 1990), EPA classified membrane filtration processes as alternative technologies, and required that pilot studies or other means be used to demonstrate that these technologies meet the requirements of the SWTR. The exception was RO, which was considered by EPA to be effective for removing Giardia cysts and viruses. Consequently, demonstration was considered unnecessary. This consideration was based primarily on the nominal cutoff of RO membranes (<0.001 microns [μm]) relative to the size of the pathogens of concern (0.025 μm to 10 μm). However, at the time of SWTR promulgation, all or nearly all RO facilities located in the U.S. were treating brackish groundwater supplies for the purpose of dissolved solids reduction (demineralization).

Pathogen removal was not a treatment requirement for these plants. Consequently, data demonstrating RO removal capability was not developed.

The increasing use of high-pressure membranes to purify municipal wastewaters to a quality suitable for indirect potable reuse has placed increased emphasis on the their ability to remove both protozoan cysts/oocysts and viruses. In some instances, removal efficiency required by high-pressure membranes for Giardia, Cryptosporidium and viruses is significantly greater (5 logs or higher) than what is required under the SWTR or interim ESWTR because wastewaters contain orders of magnitude greater concentrations of these pathogens.

Given the development of the these regulations and the tendency toward increasing stringency in the control of acute health risk associated with waterborne pathogens, there is a need to understand the mechanisms by which high-pressure membranes remove microorganisms of concern and, as importantly, to develop a practical monitoring tool to assess and confirm the integrity of full-scale systems that will be constructed and operated using high-pressure membranes for drinking water treatment.

BACKGROUND INFORMATION

Membrane Process Principles

Transport Equations

Lonsdale et al. (1965) proposed the following equations to represent the permeation of water and solutes across homogeneous reverse osmosis membranes:

$$J_v = A(\Delta p - \Delta \pi) \tag{1.1}$$

$$J_s = J_v c_p = B \Delta c \tag{1.2}$$

in which J_v is the volumetric solution flux (m/s), J_s is the gravimetric solute flux (mol/[m^{2}s]), A is the water transport coefficient (m/[s Pa]), B is the solute transport coefficient (m/s), c is the solute concentration (m^3/s), p is the hydraulic pressure (Pa), π is the osmotic pressure of the solution (Pa), $\Delta c = (c_w\text{-}c_p)$, $\Delta p = (p_f\text{-}p_p)$, and $\Delta\pi = (\pi_w\text{-}\pi_p)$. The subscripts p, f, and w refer to permeate, bulk feedwater, and feedwater next to the membrane wall, respectively.

Equation 1.2 can be rearranged into the following form (Urama and Mariñas, 1997):

$$\frac{c_p}{c_w - c_p} = \frac{B}{J_v} \tag{1.3}$$

In general, the feed concentration next to the membrane wall differs from the bulk feed concentration as a result of the concentration polarization phenomenon. Brian (1966) proposed the following expression to represent concentration polarization:

$$\frac{c_w - c_p}{c_f - c_p} = \exp\left(\frac{J_v}{k}\right) = \exp\left(\frac{J_v}{D/\delta}\right) \tag{1.4}$$

in which k is the solute mass transfer coefficient (m/s), D is the solute molecular diffusivity in water (m^2/s), and δ is the thickness of the concentration polarization layer (m).

The coefficients k and D for two separate solutes, for example, sodium chloride and sodium sulfate, can be related with the expression (Winograd et al., 1973):

$$\frac{k_{NaCl}}{k_{Na_2SO_4}} = \left(\frac{D_{NaCl}}{D_{Na_2SO_4}} \right)^{\beta} \tag{1.5}$$

with β (dimensionless) equal to 2/3 for a clean membrane surface or 1 for a fouled membrane (Urama and Mariñas, 1997).

Equations 1.3 and 1.4 can be combined to give the following expression (Urama and Mariñas, 1997):

$$\frac{c_p}{c_f - c_p} = \frac{B}{J_v} \exp\left(\frac{J_v}{k} \right) \tag{1.6}$$

Equation 1.6 can be used to fit sets of experimental data (c_f, c_p, J_v) by for example least-squares regression to determine corresponding B and k values. Equation 1.6 simplifies to Equation 1.3 with $c_w=c_f$, if concentration polarization is negligible (i.e., $k \approx \infty$). In such case, a plot of $c_p/(c_f\text{-}c_p)$ versus $1/J_v$ should result in a straight line having a slope of B, and passing through the origin.

Assessment of Solute Leakage

A method to quantify the amount of solute passing across the membranes through imperfections (i.e., solute leakage) was proposed by Urama and Mariñas (1997). Equation 1.2 was modified to include a term representing the leakage of solute directly into the permeate:

$$J_s = J_v c_p = B\Delta c + \varepsilon J_v c_w \tag{1.7}$$

where ε represents the fraction of total permeate passing the membrane through imperfections. Equation 1.7 assumes the concentration of the solute passing the membrane through imperfections is the same as the concentration of solute at the wall of the membrane (c_w). Equations 1.4 and 1.7 can be combined to develop a relationship between feed and permeate solute concentrations in terms of ε, B, J_v, and k when permeation through imperfections is a significant component of solute transport. The resulting expression is:

$$\frac{c_p}{c_f - c_p} = \left(\frac{\varepsilon}{1-\varepsilon} + \frac{B}{(1-\varepsilon)J_v} \right) \exp\left(\frac{J_v}{k} \right) \tag{1.8}$$

In order to obtain an estimate for ε using Equation 1.8, either B or k must be known to obtain a statistically significant fit using nonlinear least-squares regression. Urama and Mariñas (1997) proposed using both highly rejected and weakly rejected solute data to obtain an estimate for ε. Equation 1.6 can be used to model the weakly rejected solute data since a small leakage contributes only a small percentage of total solute passage for a relatively weakly rejected solute. Once B and k have been determined for the weakly rejected solute, Equation 1.5 can be used to determine the k for the highly rejected solute based on the ratio of diffusivities of the two solutes with $\beta = 1$ in the more common case of the presence of a fouling film. Knowing k for the highly rejected solute, Equation 1.8 can then be used to fit the highly rejected solute data to find B and ε.

Membrane Integrity Studies

Sorber, Malina, and Sagik (1972) conducted the first reported study of high-pressure membranes for removal of viruses. Asymmetric UF and RO cellulose acetate (CA) membranes, ranging in conductivity rejection from 35 to 97 percent, were employed in coupons of 2 square inches (in^2) (12.9 square centimeters [cm^2]) active filtering area. The four membrane types were challenged with both Poliovirus and Coliphage T2, at feedwater concentrations ranging from 9.6×10^1 to 6.8×10^8 plaque forming units (pfu) per liter (L). Although small concentrations of viruses were detected in the product water, removal efficiency of greater than 5 logs (>99.999%) was obtained where feedwater titers exceeded 10^6 pfu/L. Virus removal efficiency was independent of membrane type (i.e., the lowest rejecting UF membrane provided as high a removal efficiency as the highest rejecting RO membrane) and membrane flux (similar concentrations of virus were observed in permeates from low flux RO and high flux UF membranes). The researchers concluded that the presence of viruses in the product water was caused by random penetration of virus particles through defects in the rejecting surface of the cellulose acetate membrane. At virus concentrations naturally occurring in raw water supplies, such random penetration would be considered highly unlikely.

The need to achieve pathogen removal with high-pressure (HP) membrane processes was initially driven by public health regulations associated with the reclamation and reuse of municipal wastewaters. Introducing these reclaimed waters into potable water aquifers through direct injection required that advanced treatment provide a very high level of virus removal. A 1979 study conducted at the Orange County Water District (OCWD) Water Factory 21 in southern California on the facility's full-scale process train of lime treatment, recarbonation, granular media filtration, and RO showed complete virus removal (OCWD and James M. Montgomery Consulting Engineers, Inc., 1979). The study evaluated removal of endemic viral populations; no measurements were made of virus removal through the RO process alone.

More recently, a number of researchers have performed challenge studies on RO pilot systems using both protozoan cysts and oocysts, and viruses in an attempt to more accurately evaluate the ability of RO membranes to retain pathogens of concern in drinking water. The MS2 phage is used routinely in challenge studies in lieu of enteric viruses because of its similarity in size, shape, and nucleic acid to polio virus and hepatitis. Again, these studies were performed in the context of the advanced reclamation of municipal wastewater where secondary and tertiary effluents represented the RO feedwater. In each study, arrays of multiple spiral-wound membrane element series arranged in housings were evaluated at pilot scale in contrast to the use of small section of flat-sheet membrane. In one study (Rose and Farrah, 1993), CA and interfacially formed composite (COMP) membranes were challenged with four types of bacteriophage associated with the following host organisms: E. Coli, Pseudomonas phaseolicola,

Salmonella typhimurium and Vibro spp and included MS2 phage. The CA membranes provided better virus removal than the COMP membrane. Average removal for the CA membrane was 4.5 logs, and ranged from 2.25 to 6 logs. Average log removal for the COMP membranes was 3.73 logs, ranging from 3.22 to 3.91 logs. Each type of RO membrane was operated at 80 percent feedwater recovery and 11 gallons per day per square foot (gfd) (18.68 liters per hour per square meter [L/m^2/hr]) flux.

A similar study (Gagliardo *et al.*, 1997a) challenged CA and COMP membrane with *Giardia lamblia*, *Cryptosporidium parvum* and MS2 *phage*. Both membrane systems were operated at 50 percent feedwater recovery and flux of 7 to 10 gfd (11.9 to 17.0 L/m^2/hr). The RO concentrate was recycled to the RO feedwater to simulate higher recovery. Initial tests showed protozoan removal to be <5 logs and virus removal <3.3 logs. Protozoan removals improved to >5.7 logs following detection and replacement of faulty product water tube o-rings (no cysts/oocysts detected in the product water). Although o-ring replacement failed to improve virus rejection for the COMP membrane, it did increase CA membrane removal to 5.1 logs. The authors concluded that other factors, such as the membrane itself or glue lines sealing the edges of the membrane leaves, may be more critical to the removal of viruses.

A follow-on study (Gagliardo *et al.,* 1997b) used the same challenge microorganisms and four different manufacturer's COMP membranes (including the one used in the prior study). All membranes exhibited complete removal of protozoa (>4.8 logs). The MS2 *phage* removal ranged from 2 to >5.9 logs. The results are surprising in that the same polymer (aromatic polyamide) is used to form the semi-permeable rejecting layer in each manufacturer's membrane. No correlation was evident between the rated sodium chloride rejection of the four membranes and observed virus removal. In fact, the membrane having the highest salt rejection rating showed the lowest virus removal.

In an attempt to elucidate the mechanism for the variable virus passage in the two studies conducted by Gagliardo *et al.*, (1997a,b), additional studies were performed using flat-sheet membrane sections (Adham *et al.*, 1997,1998a) and spiral-wound elements (Adham *et al.*, 1998b). The approach was to evaluate virus rejection by the RO membrane structure without the impact of potential leakage through defective product-water tube seals or membrane leaf glue lines. These authors used MS2 bacterial *phage* viruses as surrogate indicators for enteric viruses. Five RO membrane coupons were tested at the experimental conditions outlined in Table 1.1. Flat leaf tests were performed with the five RO membranes listed in Table 1.1. Experiments were conducted using a 12-inch in diameter circular cell with no concentrate return line, i.e., dead-end system. Target feed concentrations ranged from 10^3 to 10^8 pfu/ml. Each test was conducted by pressurizing the feed solution at 120 psi and collecting sample until 5 L of permeate was processed. Experimental results for the flat-leaf tests are summarized in Table 1.2.

Two of the three composite membranes (B and C), as well as one of the two cellulose acetate membranes (E) allowed MS2 *phage* to pass from the feed into the permeate. Adham *et al.* (1998a) noted that the experiments were conducted in a dead-end configuration, whereas, cross-flow configurations are typical in actual treatment plants. A cross-flow configuration may allow for additional log removals due to shear forces acting at the surface of the membrane, which could prevent virus accumulation. Also, MS2 *phage* experiments were conducted in particle-free feedwater. In a typical feedwater observed in the field, viruses may attach to particles present in the feedwater, which may enhance MS2 removal.

In addition to flat-leaf MS2 *phage* testing, Adham *et al.* (1998b) also conducted spiral-wound experiments using some of the RO membranes previously tested in flat-leaf configuration. In general, membranes that allowed virus passage in the flat-leaf configuration also allowed virus passage in the spiral-wound configuration. Table 1.3 summarizes spiral-wound virus data reported by Adham *et al.* (1998b).

Table 1.1
Characteristics of flat-leaf membranes tested by Adham et al. (1998a)

Parameter	Membrane				
	A	B	C	D	E
Manufacturer	FilmTec	Hydranautics	Fluid Systems	Fluid Systems	Hydranautics
Commercial Name	BW30HP	ESPA-UHA	TFCLHR	ROGAHR	Desal
Membrane Type	Reverse Osmosis	Reverse Osmosis	Reverse Osmosis	Reverse Osmosis	Reverse Osmosis
Active Layer Polymer Type	Polyamide (TFC)	Polyamide (TFC)	Polyamide (TFC)	Cellulose Acetate	Polyamide (TFC)
pH Range	2.0 – 11.0	3.0 – 10.0	4.0 – 11.0	4.0 – 6.0	2.0 – 11.0
Max. Turbidity (ntu)	1.0	1.0	1.0	1.0	1.0
Max. Silt Density Index (15 min.)	5.0	4.0	4.0	4.0	5.0
Max. Temp. (°C)	45	45	45	40	50
Free Chlorine Resist.	<0.1 ppm	<0.1 ppm	0.0 ppm	1.0 ppm	500 ppm-h
Max. Press. (psi)	600	600	600	600	500

Table 1.2
Summary of results reported by Adham et al. (1998a) for MS2 phage experiments with flat-leaf RO membrane coupons

Parameter	Membrane				
	A	B	C	D	E
Number of Coupons Tested	11	10	14	3	4
Median MS2 Log Removal	>6.5	5.6	2.7	>4.9	4.6
Range MS2 Log Removal	>3.4 – >7.4	3.7 – 6.9	1.4 – 5.8	>4.5 – >6.0	1.0 – 5.0

Table 1.3
Summary of results reported by Adham et al. (1998b) for MS2 phage experiments with spiral-wound RO membrane elements

Parameter	Membrane		
	A	B	C
Number of Elements Tested	9	9	9
Median MS2 Log Removal	>5.6	4.9	3.2
Range MS2 Log Removal	>5.3 – >5.9	4.2 – >5.6	3.1 – 3.6

The studies by Adham *et al.* (1998a,b) shown that viruses, even though larger than the typical pore size of reverse osmosis and nanofiltration membranes, are capable of permeating through certain membranes. Bench-scale experimental tasks for this study were designed to confirm these findings and to assess mechanisms responsible for virus permeation across the membranes.

Some general conclusions can be drawn from the above-cited studies:

- When product water tube seal integrity is maintained, spiral-wound RO elements provide for complete removal of protozoan cysts and oocysts (Gagliardo *et al.* 1997a,b).
- Passage of virus through high-pressure membranes appears to be independent of membrane flux and membrane salt rejection (Sorber, Malina, and Sagik 1972; Gagliardo *et. al.*, 1997a,b).
- Virus rejection is more variable for COMP membranes than for CA membranes (Adham *et. al.*, 1997).
- Virus rejection appears to be controlled by the integrity of the rejecting layer rather than by the element glue line or product-water tube seals.
- Results obtained with challenge studies using high-feed concentrations of microorganisms directly extend to rejection characteristics at lower concentrations (Sorber, Malina, and Sagik 1972). If virus passage is a random occurrence through membrane imperfections, results obtained during challenge studies may overestimate passage through high-pressure membranes when operated on natural water supplies containing considerably lower ambient virus concentrations.

MECHANISMS FOR MICROORGANISM PASSAGE

Sorber, Malina, and Sagik (1972) and Yahya *et al.* (1993) concluded that the mechanism of virus penetration of asymmetrical CA membranes is not obvious and that insignificant penetrations are most likely due to random passage though mechanical defects in the membranes or passage through very large pores resulting from random imperfections in the cross-linkages of the cellulose acetate. Rose and Farrah (1993) attributed Coliphage passage to a defective o-ring. Once the o-ring was replaced, Coliphage counts decreased to undetectable in the permeate. The membrane did not pass viruses, and was considered to serve as a barrier between the feed and permeate side of membrane. A later study by Trussell *et al.* (1995) found that replacing damaged o-rings eliminated the passage of cysts and oocysts, but not the passage of viruses. The authors concluded that other factors, such as the membrane itself or glue lines sealing the edges of the membrane leaves, might be more critical to the removal of viruses. Overall below mechanisms may be responsible for virus and cyst passage from the spiral-wound elements and the method for arranging elements in the housing:

- Imperfections in the membrane sheet (barrier layer/support layer/backing material)
- Imperfections in the glue or heat scals of the membrane leaf
- Imperfections in the membrane at the area of attachment to the product water tube
- Incomplete sealing by o-rings used to isolate the feed and permeate streams on the product-water core tube.

ASSESSMENT OF CURRENT SURROGATES

Many studies assessed the integrity of high-pressure membranes for pathogen removal by directly measuring the degree of passage of microorganisms such as *Giardia, Cryptosporidium*, or the enteric virus surrogate, MS2 *phage*. Little, if any, work has been reported in the literature on use of non-microbial surrogates. The surrogates that have been reported include turbidity, particle counts, and conductivity. The first two were originally developed for assessing the particle removal performance of conventional water treatment processes. These surrogates are inappropriate for evaluating particle removal by high-pressure membrane systems because feedwaters to these systems must, by design, have low turbidity and particle counts to operate efficiently. High-pressure membrane manufacturers typically recommend that feedwater turbidity be less than 1 nephelometric turbidity unit (NTU) and preferably $\leq$0.2 NTU (Fluid Systems 1995; FilmTec Corporation 1997). Although particle counting of RO feedwaters is only now becoming routine, it is estimated that particle counts would be <1,000/ml at these turbidity levels, limiting maximum removal measurement to <3 logs. Further, turbidity and particle counting are currently impractical for measuring particles the size of enteric viruses (<100 nanometer, nm).

Other surrogate methods that have been tested include ultraviolet light absorbance at 254 nm wavelength, commonly referred to as (UVA_{254}). The UVA_{254} method has been applied to systems using high-pressure membranes with low total dissolved solids (TDS) rejection but high rejection of organic compounds that adsorb light in this wavelength (Spangenberg 1997). UVA_{254} measurement has the potential to provide as much as 3-log removal resolution for source waters with absorbance values greater than 0.6 cm^{-1} and where long path length sample cells (10 cm) are employed to measure UVA_{254} in the permeate at a detection limit of 0.001 cm^{-1}.

One technique that can be used to assess the presence of imperfections in the high-pressure membrane structure is mixed solute testing. With this method, the membrane is challenged using high concentrations of a mixed solute solution containing a highly rejected component, such as magnesium sulfate ($MgSO_4$), and a poorly rejected component, such as a borate salt (Urama and Mariñas, 1997). Where imperfections (leaks) are present, the $MgSO_4$ passage will increase preferentially to borate passage.

Fluorescent dyes such as microspheres and Rhodamine WT (R-WT) may also be used as potential virus surrogates. Microspheres are fluorescent carboxylate-modified FluoSpheres® (Molecular Probes, Eugene, OR) that may be customized to have different molecular characteristics and size. According to the "Product Information" provided by Molecular Probes (2000), the carboxylate-modified microsphere is made by grafting polymers containing carboxylic acid groups to sulfate microspheres. The result is a microsphere with a highly charged, relatively hydrophilic, and, somewhat porous surface layer. The manufacturer also reported the surface charges of carboxylate-modified microspheres range between 0.1 and 2.0 milliequivalents/gram.

Rhodamine WT is a multi-aromatic ring organic compound having a molecular weight of 480, and is approved for use by the EPA as a tracer in water treatment plants. Characteristics of R-WT that make it attractive as a viral surrogate with high-pressure membranes include the following:

- It may have low adsorbability on solid surfaces.
- It should be highly rejected by both NF and RO membranes having molecular weight retentions of <300 daltons.
- It represents a conservative surrogate relative to enteric viruses. Although no hydraulic radius is reported for the compound, as a dissolved (non-particulate) substance, R-WT is smaller in size than both MS2 *phage* and enteric viruses. Consequently, demonstrating a high level of rejection of R-WT by RO and NF membranes should result in equal or higher demonstrated virus rejection.
- It can be quantified in clean (potable) water at very low concentrations (10 ng/L).

The R-WT dose to membrane feedwater in this project was about 1 part per million (ppm). The EPA has recommended that the concentration of R-WT be limited to 10 parts per billion (ppb) in raw waters to drinking water facilities (Turner Designs, 2000). This limit assumes no removal of the compound by conventional water treatment and the presence of 10 ppb in finished drinking water. It is reasonable to assume 1 ppm could be dosed in the feed to a RO/NF facility based on a minimum rejection of 99 percent. Such rejection would result in a permeate concentration of 10 ppb. The fluorometer used in this research (TD700 laboratory fluorometer, Turner Designs, Sunnyvale, CA) had a R-WT minimum quantification level of about 20-25 ng/L. At the feed dosage of 1 ppm, the resolution of R-WT by the fluorometer across the membrane was about 5 log, assuming complete rejection. This equals or exceeds the current EPA drinking water regulation virus removal requirement of 4 log.

The potential benefits of R-WT as a surrogate over both MS2 *phage* and fluorescent microspheres are as listed:

- R-WT is a readily available, relatively inexpensive material whose health effects through ingestion have already been evaluated by the Food and Drug Administration and EPA.
- Drinking water utilities have experience in using the compound for hydraulic tracer studies in water plants.
- Field and laboratory analytical instruments for measurement of R-WT at the necessary detection level are commercially available and affordable.

The limitations of R-WT relative to fluorescent microspheres and MS2 *phage* are as listed:

- Potentially lower membrane rejection (as a medium weight organic compound).
- Lower resolution of integrity quantification (4-5 logs at current EPA accepted feed dosage versus >6 logs for MS2 *phage*).

A summary of the methods in use to monitor membrane integrity has been developed and is shown in Table 1.4. Several indirect methods have been applied for many years to the integrity testing of low-pressure membranes (bubble point, air pressure hold, and diffusive air tests), while others (vacuum and dye tests) have been developed as ASTM (American Society for Testing and Materials) test methods to detect leaks with spiral-wound RO membranes (ASTM Standards, 1988). Although some of these integrity methods have shown some promise, they have also demonstrated limitations. A number of the integrity monitoring techniques listed in Table 1.4 are new to water treatment. However, they have been used in the pharmaceutical field, and may be good candidates for use in membrane applications.

Table 1.4
Methods for testing membrane integrity

Integrity Test	Description	Application (Membrane Type)
Bubble Point	Non-destructive check on performance. Test is performed by wetting the membrane filter with water, increasing the pressure of air upstream of the filter, and watching for air bubbles downstream. The pressure at which a steady stream of bubbles appears is the bubble point pressure. Pressures lower than bubble point indicate a damaged membrane.	Low-pressure
Diffusion	For high-volume systems where large volumes of downstream water must be displaced before bubbles can be detected. Under pressure, air flows through the water-filled pores at differential pressure below bubble point pressure by diffusion following Fick's law. Can only be measured in large area filters. The test involves applying pressure at 80 percent of bubble point. Applying pressure at 80 percent of bubble point validates filter integrity because damaged filters would be determined at lower pressures.	Low-pressure
Millipore Corrtest	A patented test by Millipore designed to verify the virus removal capabilities of the Viresolve line of membranes. This non-destructive test involves measuring the product flow of a membrane. The membrane is wetted with polyethylene glycol, and an ammonium sulfate solution is introduced. The resulting product coating the membrane has properties that, at the specified feed pressure, will pass water only through pores with an effective size greater than 28 nanometers. The log of the ratio of the flow before the chemical treatment to the flow after the treatment is a surrogate for the log removal of 28 nanometer particles. *This procedure would be difficult to implement on large-scale municipal membrane-based systems.*	Low-pressure and high-pressure
Vacuum Test for Spiral-wound Devices	This test is applicable to detect leaks in spiral-wound reverse osmosis elements when such leaks are significant enough to prevent the device from holding a vacuum. These leaks might be due to damaged membrane material, glue line failures, or leaks in o-ring seals. This test is useful as a screening procedure and is unintended for absolute verification of such leaks. The device is evaluated with one end of the permeate collection tube sealed. A vacuum gauge on the other end of the tube is observed. A rapid decay in vacuum indicates a leak. *This test is semi-quantitative and is designed for initial quality control of elements prior to operation.*	High-pressure
Dye Test for Spiral-wound and Tubular Devices	This test is appropriate for detecting leaks in spiral-wound or tubular reverse osmosis devices that are due to lack of or loss of mechanical integrity. The test consists of passing a solution of a dye that is known to be removed by the membrane. The concentration of the dye in the permeate relative to that in the feed is measured by a spectrophotometer. A dye passage greater than 0.5 percent indicates a leak. *This test is intended to check for gross leakage and, as currently designed, does not have sufficient precision (<2 logs) to accomplish the goal of ensuring necessary microbial integrity.*	High-pressure

(continued)

Table 1.4 (continued)

Integrity Test	Description	Application (Membrane Type)
Membrane Probe Test	This test is designed to check each membrane element as well as the overall system. The test was developed by Irvine Ranch Water District and can be used for any spiral-wound membrane filtration system. The indicator used for this test is UVA_{254}. The UVA_{254} concentration in the specific feedwater employed by IRWD was 0.62 cm^{-1}. Typically, membranes reject approximately 99 percent of the total UVA_{254}. A test greater than 0.01 cm^{-1} would probably indicate a leak. *This test is limited in applicability as most surface waters have much lower UVA_{254} concentrations (<0.10 cm^{-1}), reducing the practical resolution between feedwater and permeate UVA_{254} to <3 logs. Further, it cannot be utilized as an on-line monitoring method.*	Low-pressure and high-pressure
Air Pressure Hold	The test involves draining the water from the permeate side of a fully wetted membrane module. The permeate side is then pressurized with air up to 15 psi and then sealed. The rate of air pressure decay is then monitored from the permeate side for a period of 5 minutes. A decline rate greater than 1 psi every 5 minutes indicates a failed test. *This test is based the bubble-point test.*	Low-pressure
Turbidity	An on-line turbidimeter is used to monitor the turbidity of the membrane permeate. Turbidity higher than 0.1 nephelometric turbidity units (NTU) would indicate a membrane leak. *By design, feedwaters to high-pressure membranes should be <1 NTU, thus limiting the resolution (between feedwater and permeate) of this measurement to 1 log.*	Low-pressure and high-pressure
Conductivity	The conductivity of the permeate from a membrane process can be used to monitor the total dissolved solids and indicate whether a membrane breach or seal rupture has occurred. High-pressure membrane systems applicable to the treatment of surface water supplies will possess a conductivity rejection of 99.5% or less. *This limits the resolution of the test to <3 logs under the best possible conditions.*	Low-pressure and high-pressure
Particle Counting	Comparing particle counts in the membrane feedwater and permeate can determine whether the membrane performance has changed due to rupture of the membrane material or seal leak. *The low concentration of particles present in feedwaters to high-pressure membrane systems limit the feed/permeate particle differential capability of this method to ~3 logs.*	Low-pressuure and high-pressure
Particle Monitoring	This method measures the fluctuation in intensity of a narrow beam of light transmitted through a sample containing either membrane feedwater or permeate. The instrument used does not count particles, but provides an index between 0 and 9999, based on the particulate load of the water stream. No calibration is required because it is a relative measure of the water quality. *Like particle counting, the low concentration of particles present in feedwaters to high-pressure membrane systems limit the feed/permeate particle differential capability of this method to ~3 logs.*	Low-pressure and high-pressure
Challenge Testing with Viruses	A high concentration of viruses is artificially introduced (seeded) into the membrane feedwater. The concentration that passes to the permeate is determined and used to calculate the log reduction of viruses. MS2 *phage* is commonly used for this type of challenge testing. *This method is thought to best represent membrane performance, but is expensive when applied on a large-flow scale and involves introducing live, though harmless, microbes to the treatment system. In addition, this approach does not lend itself to on-line monitoring as currently there is no real-time method for measuring viruses.*	Low-pressure and high-pressure
Fluorescently Labeled Viruses	Live viruses are labeled with a fluorescent compound to increase counts resolution in the membrane permeate. *This technique might have the capability to provide on-line virus monitoring but is prohibitively expensive at this time.*	Low-pressure and high-pressure

(continued)

Table 1.4 (continued)

Integrity Test	Description	Application (Membrane Type)
Fluorescently Labeled Microspheres	Microspheres the size of viruses are "grown" in the presence of a fluorescent dye, which becomes incorporated into the microsphere's structure. The microspheres are seeded in the membrane feedwater at high concentrations, and their presence measured in the feedwater and permeate using a flow cytometer or spectrofluorometer. *This method has the best potential for providing both a high-log resolution between feed and permeate concentrations, and can be measured real-time.*	Low-pressure and high-pressure
Sonic Sensing	This test consists of monitoring the noise made by air bubbles passing through an opening in the membrane or leak through an o-ring seal. A sound wave sensor is attached to headphones and to a visual display. The sensor is manually placed at different points on the membrane module during an air-pressure hold test. A membrane breach or seal leak has probably occurred, if noise above background levels noise is heard with the headphones and observed on the display.	Low-pressure

EVALUATION CRITERIA

An ideal integrity test would assess the potential for microbial passage independently of the mechanism responsible for passage. For this project, evaluation criteria were developed to screen out test methods that are inappropriate for one reason or another and to evaluate the remainder. The criteria that have been developed include the following:

- Size relative to pathogenic microbes
- Rejection by NF and RO membranes
- On-line analytical capabilities
- Cost effectiveness
- Detectable at low or very low levels in permeate
- Non-fouling
- Approved or likely to be approved as a drinking water additive

These criteria were used for test screening, to identify data gaps associated with some of the more promising tests, and to focus our selection of test efforts.

METHOD SELECTION FOR BENCH-SCALE STUDY USE

Surrogate candidates were evaluated for integrity monitoring using the criteria listed above has been performed. A summary of this analysis is presented in Table 1.5. From the results of this analysis, the following integrity monitoring techniques were selected for further evaluation on a bench scale:

- Fluorescent Microspheres
- Mixed Solutes
- Conductivity

Based on the results obtained from bench-scale experiments as discussed in the following chapter, MS2 *phage*, 0.02 μm microspheres and R-WT (not tested in the bench-scale) were the selected surrogates for membrane integrity tests during pilot-scale portion of this project. The tested elements in the pilot-scale were spiral-wound Hydranautics Model ESPA1-4040-UHT RO and Koch Fluid Systems Model TFCS-4040S NF.

Table 1.5
Preliminary evaluation criteria for integrity methods

Integrity Surrogate	Applicable to Virus Size	Maximum Practical Challenge Resolution	Potential for NSF Approval	Detectable on Line	Fouls Membrane	Rejected by Membrane	Retained on Membrane	Cost to Implement
Fluorescent Microspheres[1]	Yes	6 Log	Yes	Yes?	Possibly	Yes	Possibly	Mod
Bi-Selective Salt Test[2]	Yes	4 Log	Yes	No?	No	Yes	No	Mod
Sulfate or Phosphate	Yes	3 Log	Yes	No?	Possibly	Yes	Possibly	Low
ASTM Dye Test (Rhodamine WT)[3]	Yes	3 Log	Yes	Yes?	Possibly	Yes	Possibly	Mod
Particle Counting	No	3 Log	Yes	Yes	Yes	Yes	Yes	Low
Particle Monitoring	No	3 Log	Yes	Yes	Yes	Yes	Yes	Low
Conductivity/Resistivity	Yes	3 Log	Yes	Yes	Yes	Yes	No	Low
Turbidity	No	2 Log	Yes	Yes	Yes	Yes	Yes	Low
Membrane Vessel Probe	Yes	2 Log	Yes	No	No	--	No	High
Vacuum Hold Test[3]	Yes	3 Log	Yes	No	No	--	No	High
Air Pressure Hold	No	5 Log	Yes	--	No	--	No	Low
Sonic Testing	No	1 Log	Yes	No	No	--	No	Low
Bubble Point	No	1 Log	Yes	No	No	--	No	Low
Diffusive Air Flow	No	6 Log	Yes	No	No	--	No	Low
CorrTest[4]	Yes	6 Log	No	No	No	--	No	High

[1] Chiou et al. (1997)
[2] Urama and Marinas (1997)
[3] Annual Book of ASTM Standards (1988)
[4] Millipore Corporation (1998)

CHAPTER 2
BENCH-SCALE TESTING

INTRODUCTION

This section presents results from the first phase of the project, which was bench- (laboratory) scale experiments conducted at the University of Illinois at Urbana-Champaign. Five membranes were investigated in flat-leaf coupon configuration, and three among them were also tested with spiral-wound elements. Three major approaches were tested for assessing membrane integrity: (1) simultaneous testing of high-rejected and low-rejected solutes, (2) use of 24-nm fluorescent-dyed polystyrene microspheres as non-biological surrogate indicators for fecal viruses, and (3) use of MS2 *phage* as biological surrogate indicator for fecal viruses. Experiments were performed with membranes as received from the manufacturers as well as membranes after attempting to compromise their integrity by chemical and physical means.

MATERIAL AND METHODS

Experimental Apparatuses

Flat-leaf Cell Unit

Experiments were performed using a closed loop system consisting of three flat-leaf cells connected in series (Figure 2.1). Both feed and permeate piping and valves were composed of stainless steel 316 in order to minimize the formation of corrosion products in the system. Approximately 30 L of solution was prepared for each experiment inside a 55-L stainless steel 316 supply tank. Feedwater pH was maintained at 6.1±0.1 using a Cole-Parmer pH/ORP controller Model 5652-10 (Cole-Parmer Instrument Company, Vernon Hills, IL) in conjunction with a Lab Pump Jr. acid pump (Fluid Metering, Inc., Oyster Bay, NY) delivering a solution of 0.01 M HCl. A steady increase of approximately 0.1 pH units over a 24-hour period was observed due to the unavoidable formation of corrosion products and the corresponding consumption of H^+ in the system. Feedwater temperature was maintained at 25.0±0.2 °C with a refrigerated recirculator (PolyScience, Niles, IL) connected to a stainless steel 316 cooling coil immersed in the feedwater tank.

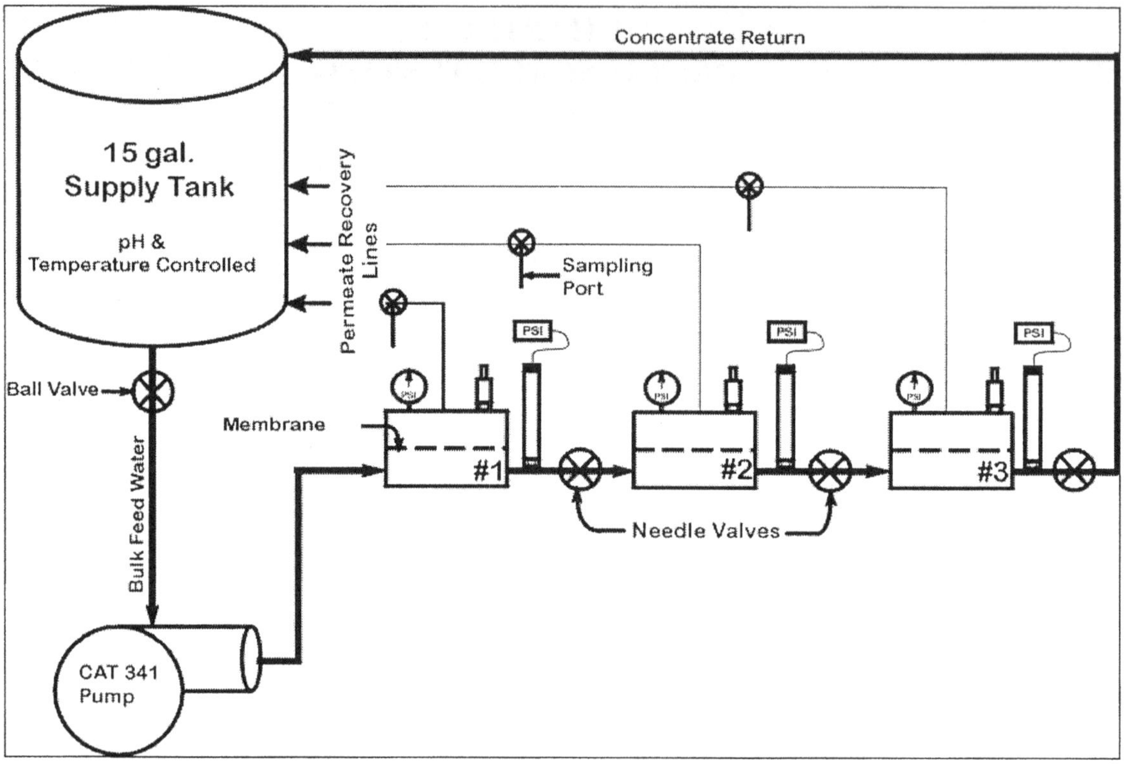

Figure 2.1 Schematic of flat-leaf experimental apparatus

A stainless steel 316 Cat pump Model 341 (Cat Pumps, Minneapolis, MN) driven by a Baldor Vector Drive Smart Motor (Baldor Electric Company, Fort Smith, Arkansas) was used to circulate the solution at 1 gallon per minute (gpm) through three SEPA CF-HP flat-leaf cells (Osmonics, Minnetonka, MN). The Baldor Vector Drive Smart Motor was capable of varying torque to the pump to maintain a steady flow, even at low rpm.

The flat-leaf cells consisted of two stainless steel 316 plates, separated by a membrane coupon (active layer facing down) and sealed by two Viton o-rings placed in the lower plate. The resulting membrane surface exposed to feed solution within each cell was 21.56 in^2. A vexor (plastic mesh) was cut from a spiral-wound element, and installed adjacent to the feed side of the membrane to promote turbulence. Permeate passing across the membrane was channeled into permeate collection lines through several small holes in the middle of the top plate. The permeate side of the membrane leaf was backed by a permeable acrylic plastic sheet located in a recessed section of the top plate that served to convey permeate from the leaf to the exit holes in the top plate. Feed pressure to each cell could be adjusted by regulating a needle valve installed in the piping downstream of each cell. Pressure was monitored with the use of electronic pressure transducers (Cole-Parmer Instrument Company, Vernon Hills, IL), Model K1, installed just upstream of the regulating valves at the exit of each cell. Pressure transducers were connected to digital readouts (Omega Engineering, Inc., Stamford, CT), Model DP205-E. Top and bottom plates of each cell were kept together and sealed with a press (Osmonics, Minnetonka, MN), pressurized to 1000 psi using either a hydraulic hand-pump (Cell #1) or pressurized nitrogen gas (Cell #2 and Cell #3).

Spiral-wound Element Unit

Spiral-wound elements were tested with the closed loop system presented in Figure 2.2. Feedwater was drawn from a high-density polyethylene tank using a constant speed pump (CAT 341, CAT Pumps USA, Minneapolis, MN), and forced through the inlet of the membrane element vessel (Model S2.5, Advanced Structures, Inc., Escondido, CA). The vessel contained a single spiral-wound membrane element, 2.5 inches in diameter and 40 inches in length. A diversion line part of the closed loop system allowed a portion of the flow exiting the pump to bypass the membrane and flow directly back into the feed tank. The amount of diverted flow was controlled with a needle valve (Whitey, Highland Heights, OH). A second needle valve was used in the concentrate return line to adjust the pressure in the element to achieve permeate flow rates within a target range. Permeate and concentrate flow rates from the element were measured with flow meters (Model RMC-141 and Model RMC-144, Dwyer Instruments, Inc., Michigan City, IN) prior to being returned to the feedwater tank. The flow meters were calibrated volumetrically. Permeate and concentrate valves (Whitey, Highland Heights, OH) were located in the corresponding return lines for sample collection.

The feed and concentrate pressures were monitored with two pressure transducers (Model 07356-50, Cole Parmer, Vernon Hills, IL) positioned at each end of the pressure vessel and connected to digital read-out devices (Model DP-25E, Omega, Stamford, CT). The feed solution pH was controlled with a system including a Cole Parmer Microprocessor Based LCD pH/ORD Controller (Model 05652-10, Vernon Hills, IL), and pH and temperature probe. The feedwater temperature was maintained constant at 25 ^{o}C by circulating water from an IsoTemp Refrigerated Circulator (Fisher Scientific, Itasca, IL) through a stainless steel 316 coil. All system components in contact with feedwater, permeate and concentrate, with the exception of the feed tank, element vessel and flow meter tubes, were made of stainless steel 316 in order to minimize the effects of corrosion.

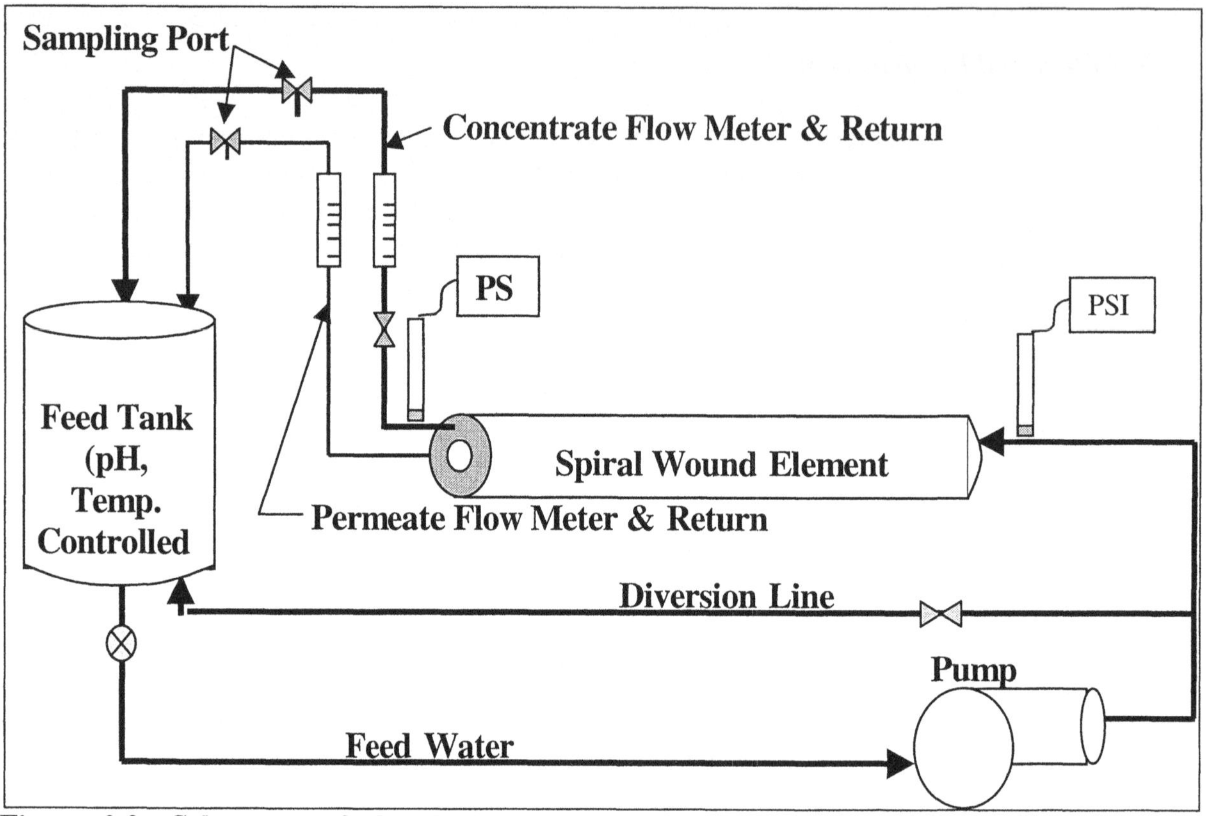

Figure 2.2 Schematic of closed loop system used for testing spiral-wound membrane elements

Chemical Solutions

Various analytical and experimental solutions mentioned in subsequent sections were prepared according to the following procedures:

1. Tri-Solute Challenge Solution: 60 g NaCl (>99.9% purity) (Fisher Scientific, Itasca, IL), 60 g Na_2SO_4 (>99.9% purity) (Fisher Scientific, Itasca, IL), 0.3 g H_3BO_3 (>99.9% purity) (Mallinckrodt Baker Inc., Paris, KY) were added to enough distilled-deionized (DDW) water to completely dissolve the constituents. The resulting solution was filtered through 0.2-μm filters to remove any extraneous particulates. The filtered concentrate solution was then added to the feed tank containing 30 L DDW producing a target concentration of 2,000 mg/L NaCl, 2,000 mg/L Na_2SO_4, and 10 mg/L H_3BO_3.
2. Carmine Reagent: Carmine reagent was prepared in accordance with Standard Method 4500B using powdered carminic acid (Aldrich Chemical Co., Milwaukee, WI) (APHA *et al.*, 1992).
3. Tween® 20 solution: 0.1 ml Tween® 20 solution (Fisher Scientific, Itasca, IL) was added to a 1L volumetric flask and brought to volume with DDW water.
4. Tryptone Yeast Extract (TYE) Broth: 1.0 g Bacto tryptone, 0.1 g Bacto yeast extract, 0.1 g Bacto dextrose (DIFCO Laboratories, Detroit, MI), 0.8 g NaCl (>99.9% purity) (Fisher Scientific, Itasca, IL), and 0.022 g $CaCl_2$ (>95% purity) (Mallinckrodt Baker Inc., Paris, KY) were added to 100 ml of deionized water. The resulting mixture was

subsequently stirred on a magnetic stir plate for approximately 10 minutes or until all constituents were completely dissolved.

5. Base Agar: 1.5 g Bacto agar were added to 100 ml of TYE broth and gently mixed. The mixture was then autoclaved for 20 minutes and placed in a 45°C water bath to cool. The agar was then dispensed in 15 ml volumes into 15 x 100mm Petri plates and allowed to harden. Once hardened, the plates were inverted and stored at 4 °C.
6. Soft Agar: 0.7 g Bacto agar was added to 100 ml of TYE broth and gently mixed. The mixture was then autoclaved for 20 minutes and placed in a 45 °C water bath to cool. The soft agar mixture was used in the MS2 *phage* assay procedure described herein.

Microspheres

A 2% solids (20,000 mg/L) stock solution of 0.02 μm orange fluorescent carboxylate-modified FluoSpheres® was obtained from Molecular Probes (Eugene, OR). The actual particle size was reported as 0.024 ± 0.005 μm. Maximum excitation/emission wavelengths were 540 nm/560 nm, respectively.

Analytical Methods

Chloride and Sulfate Ion Concentration Determination

Chloride and sulfate ion concentrations were determined in accordance with Standard Method 4110B (APHA *et al.*, 1992) using a Dionex 300 DX Ion Chromatograph (Dionex Corporation, Sunnyvale, CA). The instrument was operated with guard columns NG1 and AG9 and separation column AS9-HC (Dionex Corporation, Sunnyvale, CA). Chloride and sulfate sample concentrations were diluted into the linear range of 0.10 – 5.0 mg/L. Chloride standards of 0.25, 0.50, 0.75, 1.0, 2.5, and 5.0 mg/L, and sulfate standards of 0.1, 1.0, 2.5, and 5.0 mg/L were prepared and included with each set of samples analyzed. Typical calibration curves for chloride and sulfate ions are shown in Figures A.1 and A.2, respectively. Feed and permeate concentrations for both chloride and sulfate were obtained by interpolating between calibration curve points.

Boric Acid Concentration Determination

Boric acid concentrations were determined according to Standard Method 4500B, Carmine Method, (APHA *et al.*, 1992) using a Shimadzu UV Visible Spectrophotometer Model 1601 (Shimadzu Scientific Instruments, Columbia, MD). No dilution of permeate or feed sample was necessary due to the relatively low-feed concentration of boric acid used (10 mg/L as B). Boric acid standards of 0, 1.0, 2.5, 5.0, 7.5, and 10.0 mg/L as B were prepared and analyzed with each set of permeate and feed samples. A typical boron calibration curve is shown in Figure A.3.

Microsphere Concentration Determination

Microsphere concentrations were determined using a Bowman Series 2 Luminescence Spectrometer (SLM Aminco, Rochester, NY). An excitation wavelength of 533±4 nm or

533±8 nm (depending on concentration) and an emission wavelength of 566±8 nm were used to analyze all microsphere samples.

Although the maximum emission/excitation wavelengths were 540/560 nm, the wavelengths of 533/566 nm were used to decrease interference due to light scattering. A high-quality quartz spectrofluorometry cuvette was used in order to minimize inconsistencies caused by cuvette imperfections.

The following laboratory procedure was used to determine the concentration of microspheres in each sample.

1. The spectrofluorometer was turned on and all pertinent settings were made according to Table 2.1. The spectrofluorometer was allowed to warm up for approximately 3 hours in order to obtain consistent fluorescence readings.
2. All glassware and cuvettes to be used in the microsphere analysis were thoroughly cleaned with a 0.01% Tween® 20 solution and rinsed with DDW water in order to remove any particulates.
3. A glass vial containing a 2% solution of microspheres was sonicated for approximately 5 minutes in order to separate any clusters of microspheres that may exist in the highly concentrated stock solution.
4. Microsphere standards were prepared for the anticipated concentration range. Standards were diluted from the 2% microsphere stock previously sonicated.
5. The highest concentration standard was then pipetted into the cuvette, with care not to introduce any air bubbles. The cuvette was placed in the spectrofluorometer holder and the detector voltage was set to a level that produced a fluorescent emission of approximately 7 on the relative scale of 1 to 10. The selection of this setting ensured that all standard samples would be readable at the given spectrofluorometry settings, and that the detector would not be saturated beyond its maximum reading capacity. The orientation of the cuvette was noted and repeated for each of the subsequent samples in order to prevent any inconsistencies due to variations in the quality of the quartz on different sides of the cuvette.
6. The cuvette was removed from the spectrofluorometer and its contents emptied. The cuvette was then rinsed thoroughly with Tween® 20 and DDW water.
7. Once the cuvette was cleaned, approximately 3 ml of the next sample to be analyzed was used to rinse the cuvette and to displace any remaining DDW water that may dilute the sample. After rinsing, approximately 4 ml of sample was pipetted into the cuvette, with care not to introduce air bubbles. The cuvette was placed in the spectrofluorometer and a 30-second time trace was performed. Analysis of the time trace data is discussed below.
8. Steps 6 and 7 were repeated until all samples, including DDW water, were analyzed.

The process to obtain fluorescence values from time trace data varied with the concentration of microspheres in the sample being analyzed. High concentrations of microspheres (>50 μg/L) tended to produce very stable time trace data (Figure 2.3). Analysis of data was straightforward, and an average of 30 readings was used to determine the sample concentration. Such high concentrations of microspheres produced linear calibration curves that passed through the origin. Therefore, typically one standard sample was analyzed and used to determine the concentration of the sample from interpolation (single-point standardization).

Table 2.1
Spectrofluorometry settings used for microsphere concentration determination

Parameter	Concentration Range	
	<5 µg/L	>50 µg/L
Emission/Excitation (nm)	533/566	533/566
Bandpass (nm)	16/16	8/16
Detector Voltage (V)	860	450 – 650

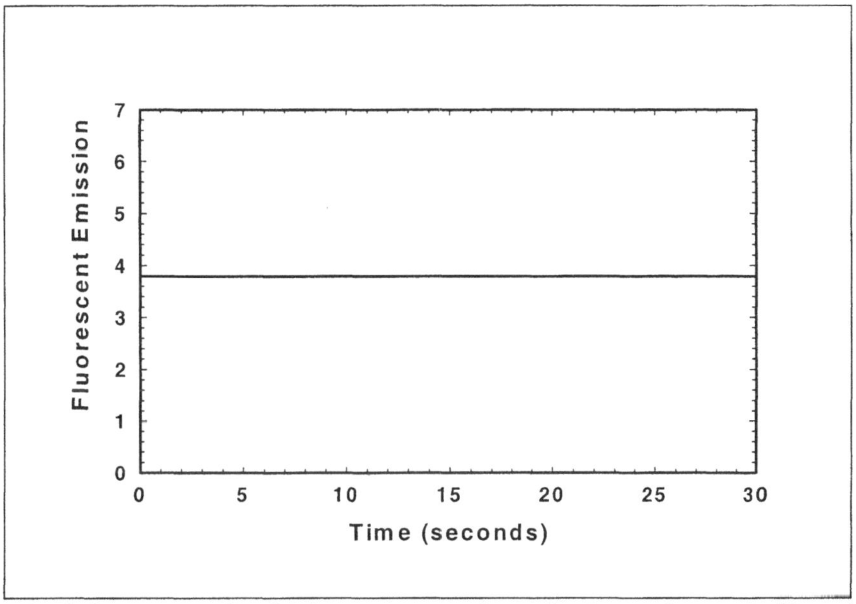

Figure 2.3 Typical time trace for high concentration of microspheres (>50 µg/L)

Samples containing 1 – 50 µg/L of microspheres produced time trace data with moderate variability (Figure 2.4). The average of 30 fluorescence measurements was also used to develop a standard curve for this microsphere concentration range. Standard curves typically contained a minimum of five concentrations. A linear regression was used instead of interpolation to calculate the concentration of microspheres in a sample from standard curve data.

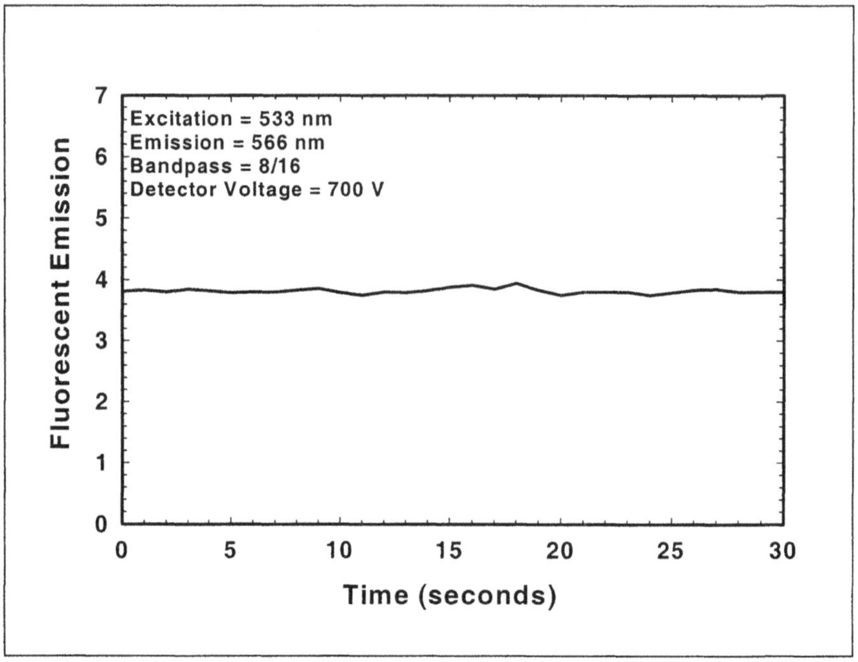

Figure 2.4 Typical time trace for medium concentration of microspheres (1–50 μg/L)

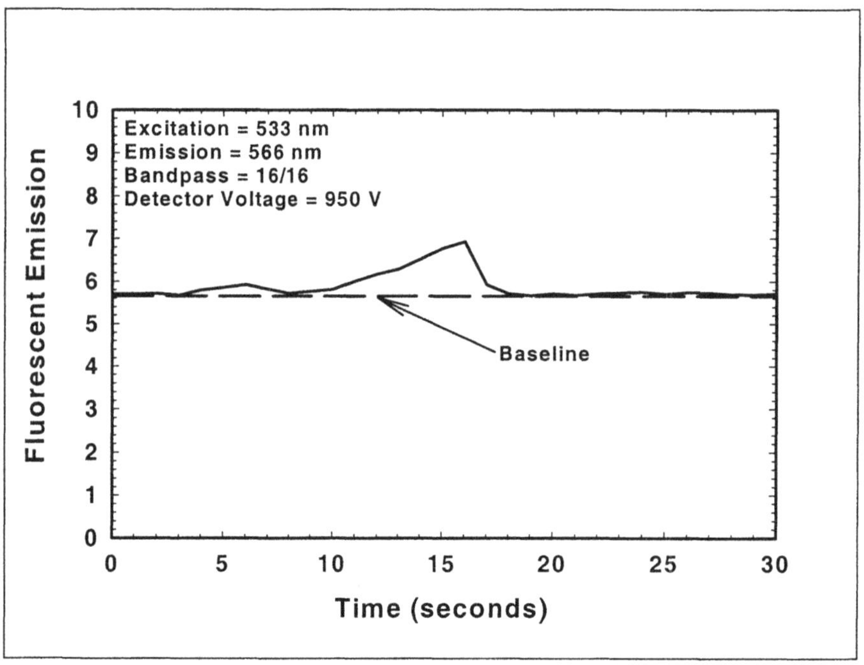

Figure 2.5 Typical time trace for low concentration of microspheres (<1 μg/L range)

Figure 2.5 shows a typical time trace at low concentrations of microspheres (ng/L range) where a large peak of fluorescence was recorded. These large interfering peaks were believed to result from the reflection of the light beam off either entrained air bubbles or particulate contamination. Despite several efforts to eliminate these peaks, their presence was more the rule than the exception. The peak in fluorescence caused the average reading of fluorescence to be skewed somewhat from the actual fluorescence of the sample. Therefore, another method of analyzing the data was investigated. This involved using the "baseline" or minimum fluorescent reading (labeled in Figure 2.5), that was assumed to be indicative of the background fluorescence of the sample. Deviations from the baseline were minimal with exception of random large peaks represented by that shown in Figure 2.5. Therefore, using the minimum fluorescent reading from all 30 points of a time trace served to filter out the effect of these large peaks. Again, all recorded fluorescent readings were used in conjunction with a standard curve calculated by linear regression of a minimum of 5 standard points to determine the concentration of microspheres in each sample.

A calibration curve representative of those used for determining the concentration of microspheres in permeate samples for both the flat-leaf cell and spiral-wound elements experiments is presented in Figure A.4 for the concentration range of 0-5 μg/L.

MS2 Phage Stock Preparation

A plaque assay method provided by the U.S. Environmental Protection Agency, Cincinnati, was used with some modifications to obtain an initial stock of MS2 *phage* virus. All tests were performed with 15597-B1 MS2 *phage* virus obtained from American Type Culture Collection (Manasass, VA), received as a freeze-dried pellet. Plates of relatively low dilution, which resulted in complete 'plaque out,' i.e., more plaques than could be counted, were prepared as described in a subsequent section, and used to produce large quantities of viruses. The top agar of these plates was harvested and placed in a tube to be compacted by centrifugation at 1900 rpm for 15 minutes. Subsequently, the supernatant, containing the *phage*, was filtered through a 0.22-μm filter to remove any remaining *E. coli* cells. The *phage* was then stored at 4 °C until addition to the system. Typically, stock concentrations ranged from 10^{11} to 10^{13} pfu/ml.

E. coli Preparation

Initially, a freeze-dried pellet of *E. coli* (ATCC, 15597) was obtained and kept at 4 °C until ready for use. Half of the pellet was dissolved in approximately 100 ml of TYE broth and incubated, under slight agitation, overnight at 37 °C. Once the initial colony of *E. coli* was prepared, the colony was kept alive through the use of hard agar slants. The hard agar slants were prepared by adding 7 ml of hard agar into 15-ml centrifuge tubes. The agar was allowed to harden, while the centrifuge tubes were tilted, thereby, increasing the surface area of the hard agar.

The hard agar slants were inoculated with *E. coli* using an inoculation loop, which was flamed before and after each inoculation. The slants were incubated overnight at 37 °C and subsequently stored at 4 °C for up to 7 days before being used to prepare *E. coli* broth.

The *E. coli* broth was prepared by introducing *E. coli* from the hard agar slants (previously stored at 4 °C) into approximately 100 ml of TYE broth and incubated, under slight agitation, overnight at 37 °C. The cycle between *E. coli* broth and slants was repeated at a minimum of every 7 days to keep the colony of *E. coli* alive.

MS2 Phage Concentration Determination

The plaque assay procedure was used to determine the titer of the prepared MS2 *phage* stock solutions as well as MS2 *phage* concentration in both feed and permeate samples. Initially, *E. coli* broth was prepared before the assay was performed. Next, the soft agar solution was prepared and allowed to cool in the 45 °C water bath. Then, feed and permeate samples were diluted, if necessary, using TYE broth in order to decrease the concentration of MS2 *phage* under 200 pfu/ml, which was the concentration of MS2 *phage* that could be easily counted on each plate. Next, a test tube rack containing 10-ml sterile glass vials with lids was introduced in the 45 °C water bath. Approximately 4 ml of soft agar was dispensed into each glass vial with care not to introduce any contaminants. Once the soft agar was added, 0.1 ml of *E. coli* broth was added to each vial. Then 1 ml of each MS2 *phage* sample was also added to individual vials. Once all of the samples were added to separate vials, the contents of the vials were emptied onto 15 x 100mm Petri dishes, containing base agar, and allowed to harden. Once the agar was firm, the plates were inverted and incubated overnight at 37 °C for a minimum of 18 hours. The plates were then removed from the incubator and the plaques were counted. The number of plaques observed on each plate was the concentration of MS2 *phage* in pfu/ml.

Several *E. coli* control samples were included with the MS2 *phage* samples in order to ensure that contamination was not a factor. For the *E. coli* controls, 1 ml of *E. coli* broth (instead of 1 ml of sample) was added with the same pipette to the control vials. The absence of MS2 *phage* in the control vials ensured the pipette was not contaminated during the assay procedure. In order to prevent contamination, all auto-pipettes were disinfected with an isopropyl alcohol solution before starting the assay. Also, between samples, after the auto-pipette tip was ejected, the manifold was cleaned with the isopropyl alcohol solution. After the assay procedure was completed, all materials used in the procedure were autoclaved in order to prevent contamination of the laboratory space.

EXPERIMENTAL DESIGN

Flat-leaf Cell Testing

Flat-leaf experiments were designed using three replicates of each membrane in order to assess performance reproducibility. Relevant properties of membranes selected for flat-leaf testing are presented in Table 2.2. Experimental conditions for each set of tests performed are summarized in Table 2.3.

Solute Experiments

Experimental sets designed to investigate the permeation of solutes across the selected membranes were performed at five hydraulic pressures. The corresponding permeate flux ranges investigated for each membrane are summarized in Table 2.3.

Table 2.2
Characteristics and performance specifications of membranes tested in flat-leaf configuration

General Information	Manufacturer	FilmTec	Fluid Systems	Hydranautics	Hydranautics	Hydranautics
	Commercial Name	BW30HP	ROGA	ESNA1	CAB2	ESPA1
	Membrane Type	Reverse Osmosis	Nanofiltration	Nanofiltration	Reverse Osmosis	Reverse Osmosis
	Active Layer	Polyamide	Cellulose Acetate	Polyamide	Cellulose Acetate	Polyamide
	pH Range	2.0 – 11.0	4.0 – 6.0	3.0 – 10.0	4.0 – 6.0	3.0 – 10.0
	Max. Press. (psi)	600	600	600	600	600
Performance Specifications	Solute / Conc. (mg/L)	NaCl / 2,000	NaCl / 2,000	NaCl / 500	NaCl / 2,000	NaCl / 1,500
	Minimum Rejection (%)	98.0	95.0	85.0	97.0	99.0
	Pressure (psi)	225	420	75	420	150
	Temperature (°C)	25	25	25	25	25
	pH Range	8.0	6.5 – 7.5	6.5 – 7.0	5.0 – 6.0	6.5 – 7.0

Table 2.3
Summary of conditions used in flat-leaf experiments

Dates	Membrane Type	Background Solute		0.02 μm Microsphere Conc. (mg/L)	MS2 virus Conc. (pfu/ml)	Permeate Flux (gfd)	Induced Imperfection
		Type	Feed Conc. (mg/L)				
6/12/98 6/21/98 6/27/98 7/28/98 8/11/98 8/25/98	BW30HP	NaCl Na_2SO_4 H_3BO_3	2,000 2,000 10	--	--	5 - 40	--
9/9/98	BW30HP	NaCl Na_2SO_4 H_3BO_3	2,000 2,000 10	10	--	3 - 30	--
9/16/98 9/23/98 10/5/98 10/12/98 10/25/98	ROGA	NaCl Na_2SO_4	2,000 2,000	--	--	5 - 25	--
11/9/98	ROGA	NaCl Na_2SO_4	2,000 2,000	--	10^8	5 - 20	--
12/7/98 12/17/98 1/18/99 1/21/99 1/22/99	ESNA1	NaCl Na_2SO_4	2,000 2,000	--	--	5 - 20	--
12/29/98 1/9/99 1/29/99	ESNA1	NaCl Na_2SO_4	2,000 2,000	--	10^8	5 - 20	--
2/8/99 2/18/99 2/22/99	CAB2	NaCl Na_2SO_4	2,000 2,000	--	10^8	5 - 25	--
3/15/99 3/16/99 3/19/99	ESPA1	NaCl Na_2SO_4	2,000 2,000	--	10^8	5-40	--
3/26/99	ESPA1	NaCl Na_2SO_4	2,000 2,000	--	10^8	30-80	Air Entrainment
5/10/99	ESPA1	NaCl Na_2SO_4	2,000 2,000	1	10^6	10-40	Scratch
5/14/99	ESPA1	NaCl Na_2SO_4	2,000 2,000	--	10^6	8-40	Scratch
5/19/99 5/21/99	ESPA1	NaCl Na_2SO_4	2,000 2,000	--	10^5	10-100	Pinhole
5/21/99	ESPA1	NaCl Na_2SO_4	2,000 2,000	0.1	--	8-90	Pinhole
5/25/99	ESPA1	NaCl Na_2SO_4	2,000 2,000	10	10^8	5-80	Pinhole

The hydraulic pressure of the system was adjusted using the needle valve at the outlet of Cell #3 (see Figure 2.1). All other needle valves were completely open in an attempt to obtain a uniform hydraulic pressure throughout the system. Slight pressure differences were observed for each cell due to head-loss in the system. Typical head-loss values of 15–20 psi between consecutive cells were observed. The system was allowed to stabilize at each target pressure for 30 minutes to 1 hour depending on the hydraulic pressure. Higher pressures produced more permeate flow, thereby, allowing the system to reach steady state faster than at lower hydraulic pressures.

After the designated equilibrium time, permeate fluxes were measured volumetrically. Three-way valves were switched to divert the flow from the permeate return lines toward sample collection ports (Figure 2.1). The flux of permeate across the membrane was calculated by dividing the volume per unit time, determined with a 10 ml graduated cylinder and a stopwatch, by the active area of membrane in each flat-leaf cell, 21.56 in^2. Following flux rate measurements, 20 ml samples were collected from each permeate line and feed tank into 40 ml glass vials for solute concentration determination.

Microsphere Experiments

Experiments were also performed with 0.02 μm orange FluoSpheres® (Molecular Probes, Eugene, OR) as non-biological surrogate indicators for enteric viruses. Microsphere testing began by adding an initial dose of 10 mg/L to the feed tank. In general, a decline in feed concentration was observed due to the microspheres adhering to the membrane surface (Figure 2.6). The concentration decreased to approximately 0.8 mg/L in a 24-hour period. The feed concentration was subsequently increased to a target of 10 mg/L twice more, and although the microsphere concentration did not stabilize, the rate of disappearance decreased with subsequent additions as illustrated in Figure 2.6. In order to account for this effect, the feed concentration was increased to 10 mg/L before each experimental set was collected for tests conducted with FilmTec BW30HP coupons. A lower target feed concentration of 1 mg/L was used for microsphere experiments with Hydranautics ESPA1 coupons. The lower concentration was a result of the refinement of the microsphere analysis procedure that allowed for additional log removals to be detected. Feed samples were collected with permeate samples at each flux in order to monitor this rate of disappearance. Experimental conditions used for microsphere testing were the same as those described previously for solute testing. Feed and permeate samples were analyzed for solute and microsphere concentration.

MS2 Phage Virus Testing

The integrity of RO and NF membranes was also assessed with MS2 *phage* viruses. MS2 *phage* was added to the feed tank through variable doses of stock solution with a target concentration of approximately 10^8 pfu/ml. The feed tank was spiked with *phage* 15 minutes prior to collecting each permeate and feed sample set. The MS2 *phage* permeate samples were collected according to the same procedure described previously for solute and microsphere testing. Sample vials were sterilized by autoclave before sampling. A time test was conducted to ensure the feed concentration of *phage* remained relatively constant during the 10 minutes after the *phage* was dosed into the feed tank. The results of this test are depicted in Figure 2.7. The concentration of the *phage* remained relatively constant during the 15-minute interval after virus introduction. Initially, approximately 5 ml of *phage* stock were added to the 30-L challenge solution to achieve a target concentration of 10^8 pfu/ml. After the initial virus introduction, subsequent doses of 1 ml *phage* stock were injected into the feed solution 15 minutes before collection of additional sample sets to counteract small drops in concentration with time.

Once virus testing for a specific membrane type was completed, the flat-leaf cells were disassembled and the coupons were replaced according to the following decontamination procedures. The pump was stopped and all piping was disconnected from the flat-leaf cells. Next, the presses were depressurized and the cells were removed. The top portion of the cell (permeate side) was removed and taken to a separate lab counter, which was previously sterilized with isopropyl alcohol solution. The separation of feed and permeate portions was a measure taken to prevent the direct contamination of the permeate portion. The membrane coupon was removed from the bottom portion of the cell (feed side) which was then drained, sterilized with isopropyl alcohol, and rinsed with deionized water. Both o-rings located in the bottom portion were removed, sterilized with isopropyl alcohol, and rinsed with deionized water before being placed back. In addition, all permeate piping was drained and disinfected by heating with a Bunsen burner.

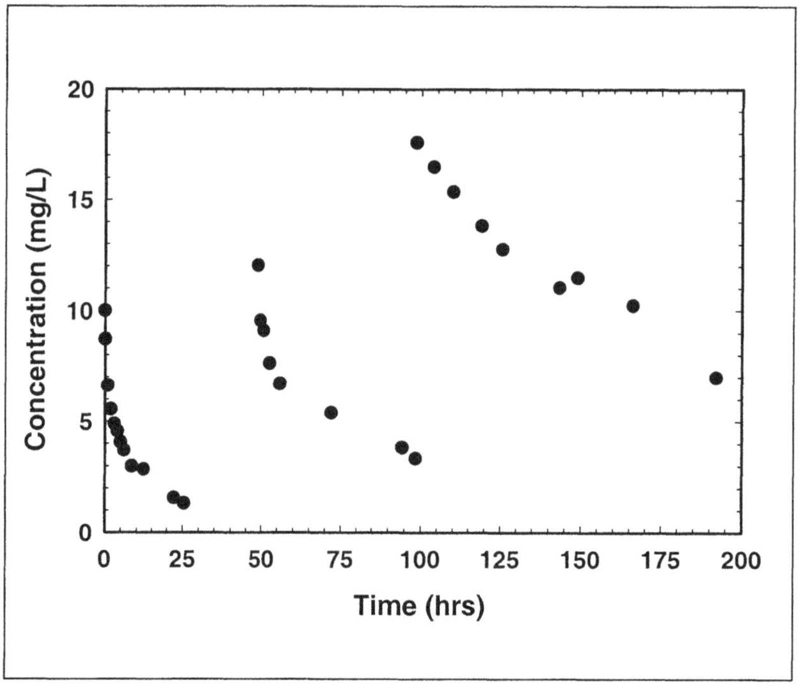

Figure 2.6 Feed concentration of 0.02-μm microspheres versus operating time from microsphere addition (BW30HP coupons)

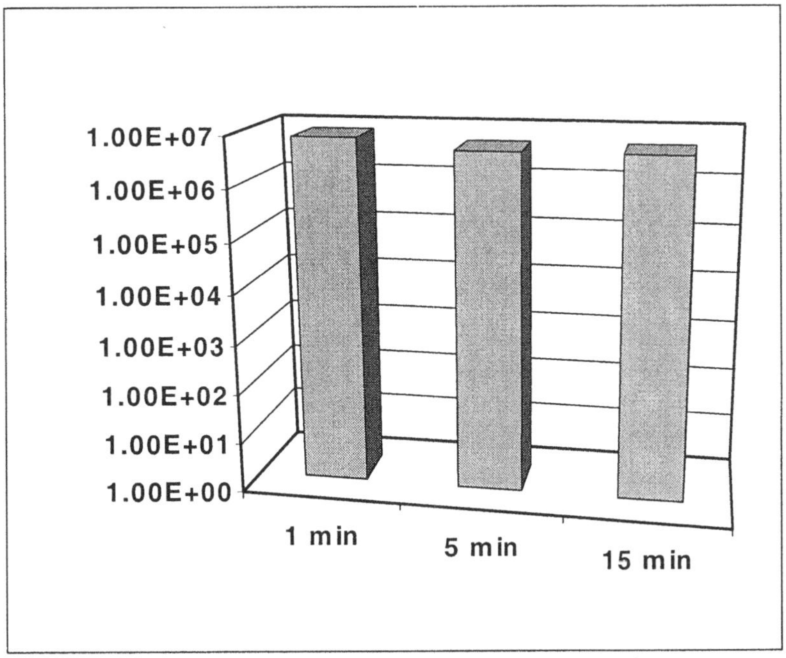

Figure 2.7 Feed concentration of MS2 phage versus time (CAB2 coupons)

Once both cell leaves were sterilized, new membrane coupons were placed on the bottom portion of the cell ensuring that the coupons covered the inner o-ring. The top leaf was then installed with care not to disturb the position of the membrane coupon on the bottom portion. The cells were then inserted into the presses, which were quickly pressurized. The piping was connected to the cells and the pump was started.

Air Entrainment Testing

Air entrainment testing was conducted in order to assess if small air bubbles that could enter or form inside the feed channel of membrane elements, subsequently attach to the membrane surface, and pass to the permeate side. In order to perform air entrainment tests, the system hydraulic pressure was set to approximately 550 psi or 50 psi less than the maximum hydraulic pressure of 600 psi specified for the Hydranautics ESPA1 coupons used. The larger the difference in feed and permeate pressure, the greater the expansion of the membrane active layer polymer caused by the attachment and permeation of expanding air bubbles. Air bubbles were introduced into the feed stream by loosening the pipe connection on the intake manifold of the pump. The ball valve between the feed tank and the pump intake was throttled back, thereby, creating negative pressure in the feed line and forcing air to enter the feed stream through the loosened connection. Air bubbles were observed exiting the concentrate return line in the feed tank confirming their presence in the feedwater. Once the presence of bubbles in the feedwater was confirmed, MS2 *phage* was seeded and samples were collected after 15 minutes. After collecting feed and permeate samples, the intake manifold connection was tightened and the feed tank valve was opened, returning the system to original operating conditions. A second set of samples was taken approximately 24 hours after operation under normal conditions in order to check if permanent membrane damage had occurred.

Imperfection Inducement

Two different types of imperfections were induced and tested using Hydranautics ESPA1 coupons. The first type of imperfection consisted of a "scratch" on the surface of the membrane that visibly removed a small portion of the active layer, but left underlying polysulfone layer intact. This imperfection was induced using a standard sewing needle by dragging the point across the ESPA1 coupons and exerting enough force to penetrate and remove the active layer.

The second type of imperfection consisted of a pinhole through both the active and polysulfone layers of the ESPA1 coupons. These imperfections were induced using the same sewing needle by forcing the needle through both layers, being careful to just barely breach the polysulfone layer. The membranes were placed active layer up on a paper towel placed on the laboratory bench. The sewing needle was then forced through the membrane, using the paper towel as a backing material. The coupon was visibly inspected by observing the amount of light passing through the imperfection as compared to the rest of the membrane. All three coupons exhibited a very small hole that was later characterized under a light microscope in order to determine the size and uniformity of the imperfections. The characterization of these imperfections is discussed in the Flat-Leaf Cell Results section.

Spiral-wound Element Experiments

Experimental Matrix

Three of the membranes investigated in flat-leaf cells were also studied in spiral-wound configuration. Manufacturers' specifications for the spiral-wound elements tested in this study are presented in Table 2.4. Summaries of experimental conditions for all tests performed with the three membrane elements investigated in this study are presented in Tables 2.5, 2.6, and 2.7 for Membranes A, B, and C, respectively. Initially, each element was tested to characterize the membrane performance with 2,000 mg/L NaCl at pH 6.0 (±0.5) and 25 °C (±1 °C), and to compare the results to manufacturers' specifications. Tests were also performed with solutions containing boric acid, 0.02-μm microspheres and MS2 *phage* viruses in addition to the 2,000 mg/L of NaCl. The purpose of the boric acid testing was to compare the rejection of a weakly (H_3BO_3) and strongly (NaCl) rejected electrolyte which could, in turn, give an indication of the integrity of the membranes. The boric acid testing, however, was abandoned after the experiments completed on the first element gave inconsistent analytical results.

After the primary characterization was complete, MS2 *phage* viruses were added to the feed tank. It was suspected that viruses would not reach a steady state concentration in the feedwater quickly. To determine dynamics of the virus feed concentration, time tests were performed. The feed tank was seeded with a stock solution of approximately 10^8 viruses/ml and samples were taken over the course of 1 hour. The time testing was performed at a constant flow rate to rule out differences in virus passage based on flow.

Once the dynamics of the virus feed concentration were determined for a given membrane, the passage of the MS2 *phage* at various flow rates was examined. If passage was observed, then further testing was performed by adding the surrogate indicator (fluorescent-dyed polystyrene microspheres) to the feed solution. These experiments would show the relationship between the surrogate indicator and the actual microorganisms. Tables 2.5, 2.6, and 2.7 list the concentrations of the various constituents that were tested with the three membrane elements.

Table 2.4
Manufacturers' specifications for spiral-wound membrane elements tested

	Membrane		
Parameter	A	B	C
Model	BW30HP[a]	ESPA1[b]	ROGA[c]
Nominal Diameter, (in.)	2.5	2.5	2.5
Membrane Active Area, (ft.^{2})	26.4	28	27
Feed Concentration (mg/L)	2,000 (NaCl)	1,500 (NaCl)	700 (TDS)
Feed Pressure (psi)	225	150	200
Feed pH	8	6.5-7.0	7.0
Temperature (°C)	25	25	25
Water Productivity, (gal/d)	650	750	260
Product Water Recovery, %	15	10	15
Nominal Solute Rejection, %	99	-	95
Minimum Solute Rejection, %	98	98	-

[a] Manufactured by Dow Chemical, Minneapolis, MN

[b] Manufactured by Hydranautics, Oceanside, CA

[c] Manufactured by Koch Fluid Systems, San Diego, CA

Testing was also done with membrane elements of compromised integrity. Chemical and physical approaches for membrane integrity compromising were used. The FilmTec BW30HP membrane was exposed to 1,000 (mg/L)-hour of free chlorine (pH 9), the upper limit specified by the manufacturer. The feedwater was circulated without applying pressure for the duration of the chlorine exposure test. The Fluid Systems ROGA membrane element was modified by inducing a relatively small pinhole and subsequently a larger perforation. The tape-wrapped element was unwrapped and the active surface of the membrane exposed for this purpose. Once the membrane was opened, a small pin was used to prick the surface of the membrane just barely passing through to the polysulfone layer. Finally, a larger perforation was induced into the membrane surface. After each imperfection was created, the membranes were re-wrapped and tested with feed solutions containing sodium chloride, *phage* and microspheres.

Table 2.5
Operating conditions and feed solution composition for experiments performed with FilmTec FT30 spiral-wound element

Exp.	pH	Temp (°C)	NaCl (mg/L)	H_3BO_3 (as mg/L B)	MS2 *Phage* Virus (#/ml)	Microspheres (mg/L)	Permeate Flux (gfd)
1	6.04	25.1	1,990	10.2	-	-	5.3-25.2
2	5.75	24.9	1,980	10.5	-	-	5.6-24.9
3	6.13	25.1	1,960	-	-	-	5.4-27.4
4[a]	6.04	25.3	1,850	-	4.55×10^7	-	8.1-36.7
5[a]	6.05	25.2	1,810	-	-	-	7.8-36.7

[a] Experiments conducted after exposing the membrane element to free chlorine for approximately 1,000 ppm-h.

Table 2.6
Operating conditions and feed solution composition for experiments performed with Hydranautics ESPA1 spiral-wound element

Exp.	pH	Temp (°C)	NaCl (mg/L)	H_3BO_3 (as mg/L B)	MS2 *Phage* Virus (#/ml)	Microspheres (mg/L)	Permeate Flux (gfd)
1	5.58	25.2	1,810	-	-	-	6.8-28.7
2	5.88	26.3	1,810	-	-	-	8.7-28.6
3	6.23	24.5	N/A	-	5.8×10^7	-	6.2-28.6
4	6.14	24.5	N/A	-	1.2×10^9	10.3	6.5-28.1
5[a]	6.20	23.8	N/A	-	2.2×10^8	7.1	6.5-26.1

N/A: Solute testing was not completed for this experiment.
[a] Experiments conducted with after compromising the o-ring sealing the permeate pipe connection to the RO vessel end-cap.

Table 2.7
Operating conditions and feed solution composition for experiments performed with Fluid Systems ROGA-HR spiral-wound element

Exp.	pH	Temp (°C)	NaCl (mg/L)	H_3BO_3 (as mg/L B)	MS2 *Phage* Virus (#/ml)	Microspheres (mg/L)	Permeate Flux (gfd)
1	6.06	25.4	1,860	-	3.6×10^4	-	9.3-22.8
2	5.94	24.9	1,710	-	1.3×10^5	-	6.7-18.2
3	6.16	25.0	1,700	-	5.5×10^6	-	6.7-17.9
4[a]	6.16	24.7	1,770	-	6.3×10^8	10.4	6.7-19.4
5[a]	6.03	24.5	1,850	-	4.1×10^8	9.9	6.7-19.9
6[b]	6.13	24.2	1,850	-	3.2×10^8	8.5	6.5-21.1

[a] Compromised element with a small pin-sized hole.

[b] Compromised element with a large perforation.

Sampling Protocol

Solute. A full solute data set included samples taken at four or five different hydraulic pressures that provided fluxes within the ranges indicated in Tables 2.5, 2.6, and 2.7. The membrane element was first tested at the lowest overall pressure, and then the pressure was gradually increased. After the highest-pressure sample was taken, the membrane was tested at the lowest pressure again to confirm that the membrane performance did not change. Table 2.8 shows the approximate time that elapsed between sample collection in the characterization testing. Usually, a 20 ml sample was taken from the feed, concentrate, and permeate, separately for solute analysis and placed into a glass or plastic vial. Glass sample vials (pre-sterilized) were used when microorganisms or surrogate indicator testing was taking place at the same time. Permeate and concentrate samples were taken from the sampling ports shown in Figure 2.2. Permeate and concentrate lines were flushed prior to sampling within an experimental set to minimize contamination from previously taken samples. The feed samples were extracted from the feed tank using a pipetter and the sampling was always done from a region of the tank that was well mixed (i.e., away from inlet and outlet). Once the samples were taken, distilled-deionized (DDW) water was added back to the feed tank to replace the volume extracted in the permeate and the concentrate samples an to correct for evaporation loses (roughly 40 ml).

Viruses. The virus sampling was conducted using almost the same procedure as the solute testing. The addition of the stock solution prior to sampling was done with a pipetter containing a clean pipette tip. The virus samples were taken approximately 5-10 minutes after the addition of the stock solution during the pressure variance testing. For dynamic tests designed to assess the variability in the feed concentration of viruses, the stock solution was only added at the beginning of the experimental set. The actual sampling procedure was identical to that of the solute with two exceptions. The glass sample vials were cleaned and sterilized before use. At the completion of the testing, the samples were stored at 4 °C until analyzed. Decontamination protocols similar to those described for the flat-leaf cell testing were also followed when switching elements.

Microspheres. Sampling of microspheres was conducted similarly to the solute and virus testing and usually the testing of the two types of particles was concurrent. Like the viruses, the microsphere were added prior to sampling with a pipetter and the samples were taken within 5 minutes after the addition of the stock solution.

Table 2.8
Typical sampling time intervals for pressure variance testing

Sample	Time, hours
1	3
2	2
3	1
4	1
5	1
6	2

RESULTS AND DISCUSSIONS

Flat-leaf Cell Experiments

BW30HP - Solute Experiments

Solute experiments with the BW30HP were performed using a tri-solute solution consisting of 2,000 mg/L NaCl, 2,000 mg/L Na_2SO_4, and 10 mg/L H_3BO_3. Multiple sets of solute data were collected for the same set of coupons, however, only two representative sets of solute data are discussed in this report. The plots in Figures 2.8-2.16 show the relationship between flux and solute passage in terms of the inverse of the flux, $1/J_v$, against the solute concentration ratio, $c_p/(c_f\text{-}c_p)$. The data set labeled 6/12/98 was collected before the addition of microspheres to the system. The data set labeled 9/8/98 was collected after adding microspheres to the system.

The experimental results shown in Figures 2.8–2.16 were first fitted to Equation 1.6, which does not take into account passage through imperfections. A nonlinear least-squares fit was performed for each data set per solute to determine *B* and *k* parameters. The resulting fitted curves are plotted in Figures 2.8–2.16 as short dashed (6/12/98) and dotted (9/8/98) lines and labeled "Regression."

The solid lines represent an attempt to fit the 6/12/98 data to Equation 1.8, which incorporates an additional parameter (ε), accounting for permeate passage through membrane imperfections by advection. Both chloride and sulfate data were fitted independently using a *k* value obtained from boric acid data combined from all three cells as described subsequently. The boric acid data was quite consistent between cells and therefore the data for all three cells were fitted simultaneously to obtain the corresponding *B* and *k* parameters. The resulting values were B = 0.359 m/d and k = 6.01 m/d. According to Urama and Mariñas (1997), the ratio of *k* values for two separate solutes should be proportional to the ratio of diffusivities of these solutes (Equation 1.5) at the level of fouling observed in this study. Therefore, the *k* values determined from the compiled boric acid data were adjusted, assuming β=1, according to diffusivities for NaCl of 1.49×10^{-5} cm^2/s, Na_2SO_4 of 1.05×10^{-5} cm^2/s, and H_3BO_3 of 1.10×10^{-5} cm^2/s. With the *k* values for each solute previously determined, Equation 1.8 was then used in a nonlinear least-squares fitting scheme to obtain ε and *B* values for both NaCl and Na_2SO_4. The resulting *B*, *k*, and ε parameters are presented in the center column under each corresponding plot in Figure 2.8 to 2.16.

The model incorporating ε did not explain the rejection of NaCl and Na_2SO_4 in a unified manner. The model predicted ε values for NaCl between 3.02×10^{-3} and 4.50×10^{-3} and ε values for Na_2SO_4 of 0 for Cells #1 and #2 and 5.77×10^{-4} for Cell #3. The ε values for NaCl and Na_2SO_4 were expected to be the same, if some small percentage of permeate was passing through the membrane by advection through imperfections. Different ε values for NaCl and Na_2SO_4 suggested that the imperfections in the membrane had selection capability favoring one solute over another. At this point, the selection capability of imperfections was not considered a feasible mechanism and the model was rejected. The addition of the ε term only added another "fitting" parameter that might not represent a valid mechanism, but allowed an additional degree of freedom and resulted in an apparently better fit.

Several other modeling attempts were made to explain the relationship between NaCl and Na_2SO_4 rejections, all of which were unsuccessful. One possible explanation for the failed modeling attempts would be the development of a dynamic membrane by a corrosion layer formed on the surface of the membranes. If a dynamic membrane were to develop, the rejection of NaCl and Na_2SO_4 might be affected differently, thus, explaining the deviations from the current model. An attempt to model a dynamic membrane was not made due to the inherent complexity of the system coupled with the limited scope of the project.

BW30HP - Microsphere Experiments

Once reproducibility was observed with the tri-solute solution, 0.02 μm microspheres were added to the system. Additional experiments were performed, analyzing permeate samples for both microsphere and tri-solute constituent concentrations. Figures 2.8–2.16 exhibit the effect the 0.02 μm microspheres inflicted on solute passage. All three flat-leaf cells showed signs of increased concentration polarization and as a result, increased solute passage.

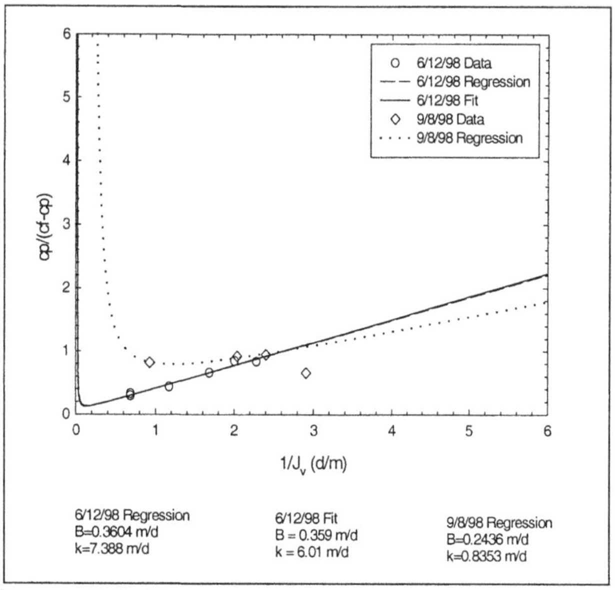

Figure 2.8 Experimental and model boric acid passage for BW30HP Cell #1

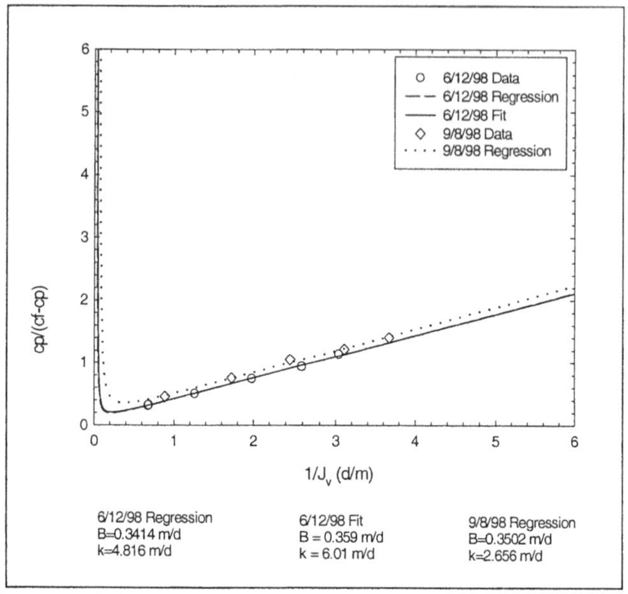

Figure 2.9 Experimental and model boric acid passage for BW30HP Cell #2

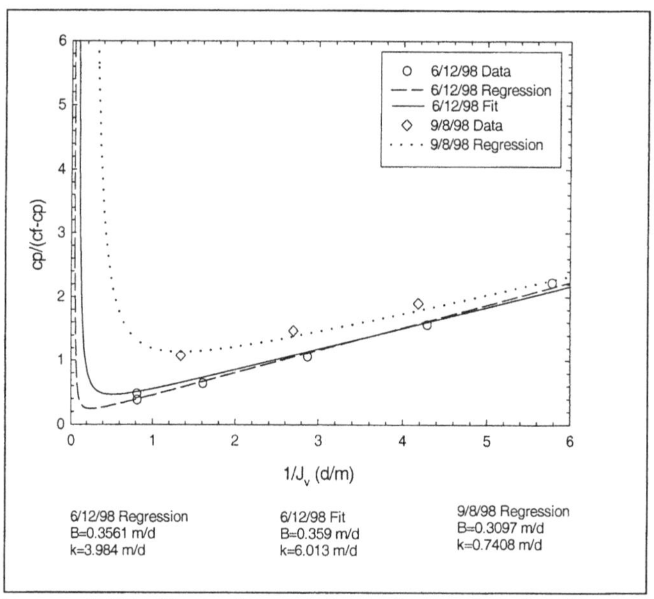

Figure 2.10 Experimental and model boric acid passage for BW30HP Cell #3

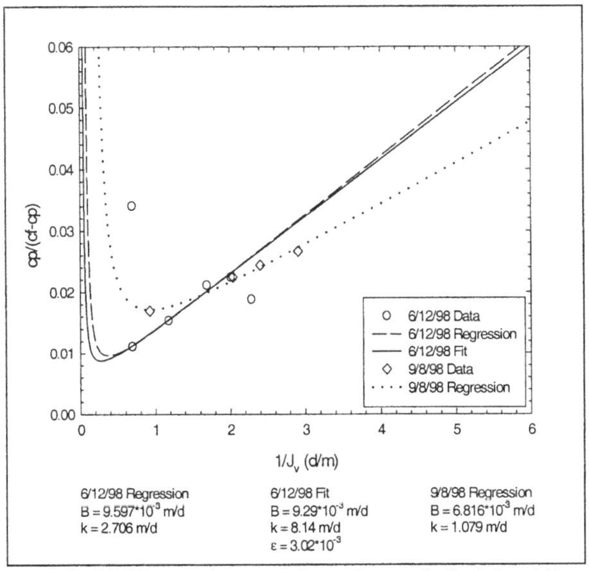

Figure 2.11 Experimental and model chloride ion passage for BW30HP Cell #1

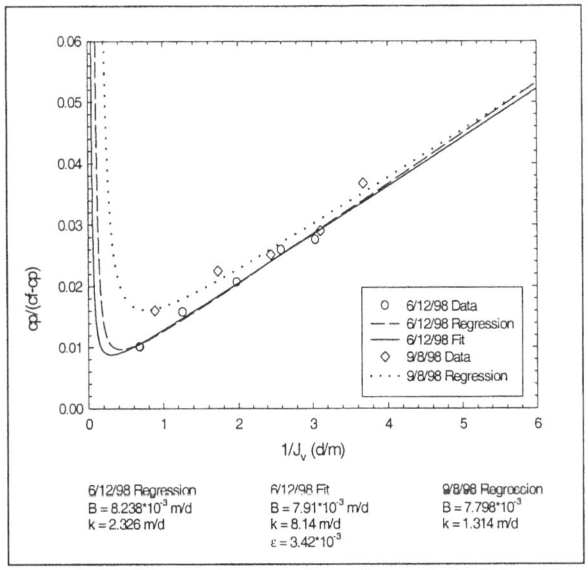

Figure 2.12 Experimental and model chloride ion passage for BW30HP Cell #2

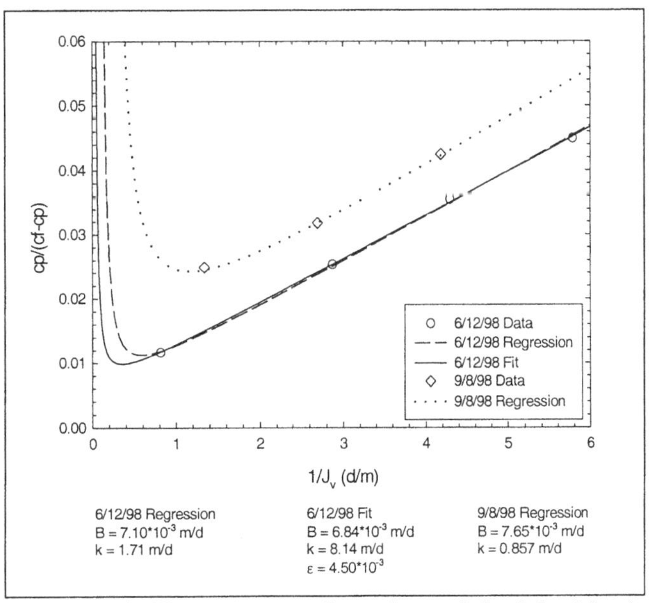

Figure 2.13 Experimental and model chloride ion passage for BW30HP Cell #3

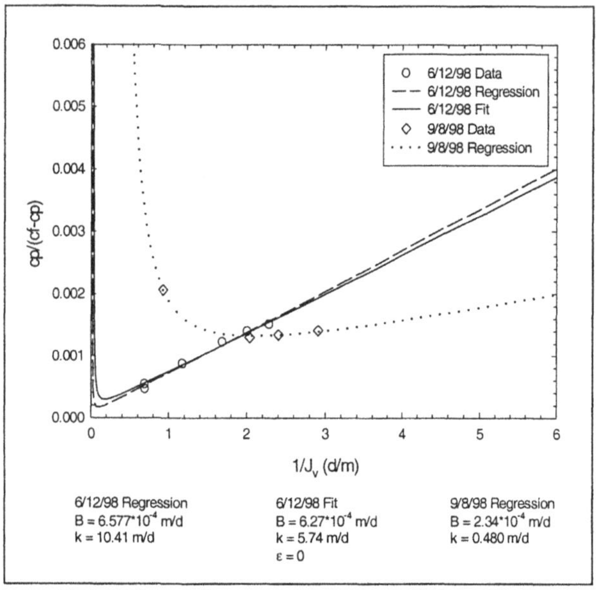

Figure 2.14 Experimental and model sulfate ion passage for BW30HP Cell #1

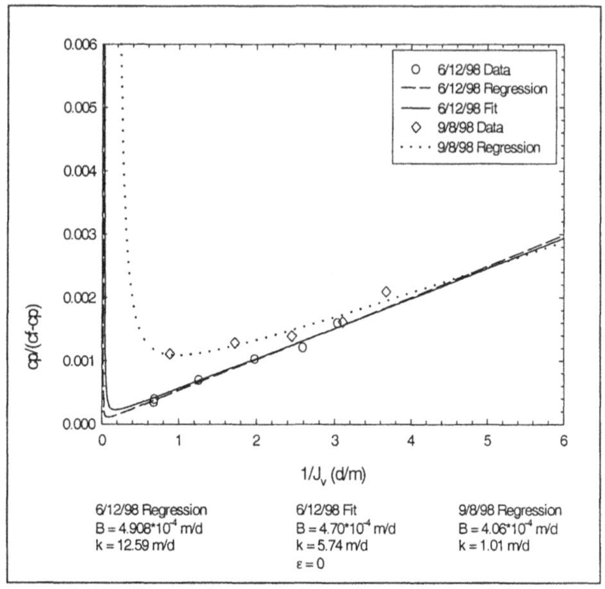

Figure 2.15 Experimental and model sulfate ion passage for BW30HP Cell #2

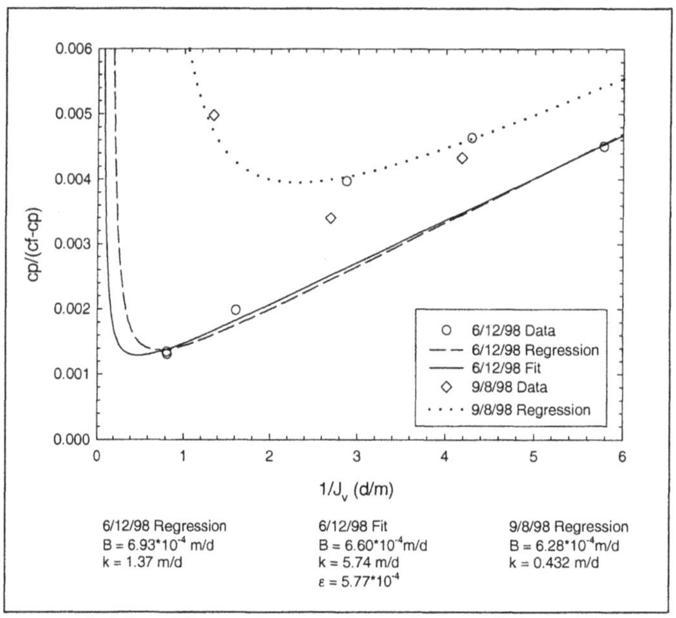

Figure 2.16 Experimental and model sulfate ion passage for BW30HP Cell #3

The results presented in Figures 2.11–2.13 show chloride ion rejections of 96 to >99 percent before microspheres were added and 92 to 98.5 percent after microspheres addition. Figures 2.14–2.16 show sulfate ion rejections of 99.5 to >99.9 percent in the absence of microspheres and 99.3 to 99.8 percent in the presence microspheres. Figures 2.8–2.10 show a smaller difference in the range of boric acid rejections (~40 to 60 percent) before and after microspheres were added because of the lower levels of concentration polarization expected for this solute due to its relatively low rejection.

The effect of increased concentration polarization caused by microsphere addition is apparent in each of the figures. A decrease in k resulted after microspheres were added to the system. Since k is the diffusivity of a given solute divided by the thickness of the concentration polarization layer, a decrease in k is the direct result of an increase in the concentration polarization layer thickness since diffusivity remains constant. This was confirmed once the membranes were removed from the system and a cake layer of microspheres was observed on membrane surfaces.

Permeate sample analyses revealed no conclusive evidence that microspheres were passing through the BW30HP coupons. The standard curve used to evaluate the concentration of microspheres in the permeate is shown in Figure 2.17. At the time microsphere samples for the BW30HP coupons were being analyzed, a significant level of laboratory contamination may have occurred, which was discovered and corrected at a later date. This may explain the 3 outliers (25 ng/L, 100 ng/L, and 250 ng/L), which were not used in the calculation of the calibration curve. The lowest concentration measured was 25 ng/L, which corresponded to ~ 5.5 log removal with respect to a target feed concentration of 10 mg/L. This was the minimum concentration that could be consistently measured with the spectrofluorometric technique used. The data collected on BW30HP permeate samples for the experimental set 9/8/98 is shown in Table 2.9.

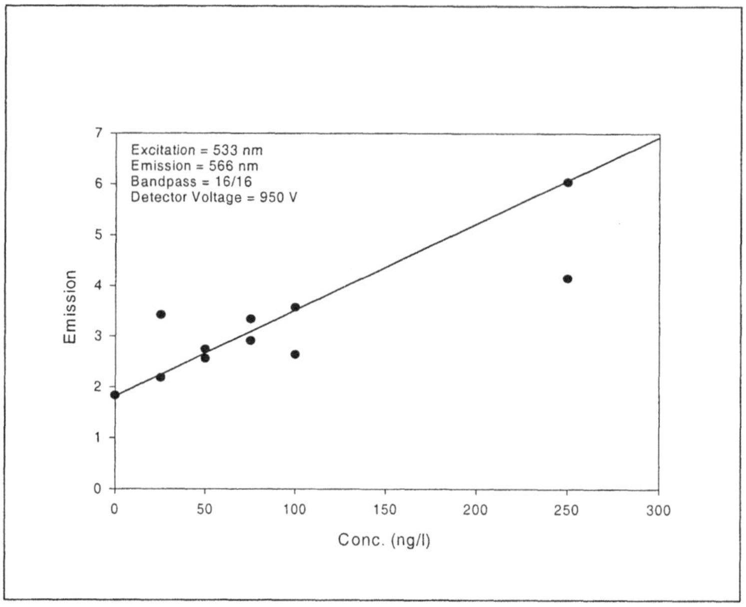

Figure 2.17 0.02 μm microsphere calibration curve for experiments performed with the BW30HP flat-leaf coupons

Table 2.9
Microsphere experimental results for the BW30HP membrane (9/8/98)

Sample	Pressure (psi)	Flux (m/d)	Permeate Emission	Concentration	
				Permeate (ng/L)	Feed (mg/L)
Cell #1	135	0.344	2.37	32	3.82
	150	0.416	2.50	40	2.90
	170	0.491	2.11	<25	2.21
	220	--	2.32	29	1.69
	347	1.08	2.66	49	<0.50
Cell #2	98	0.272	1.93	<25	3.82
	112	0.322	1.95	<25	2.90
	134	0.408	2.06	<25	2.21
	185	0.580	1.97	<25	1.69
	314	1.13	1.83	<25	<0.50
Cell #3	80	0.117	4.39	150	3.82
	94	0.163	3.74	110	2.90
	115	0.239	2.46	38	2.21
	167	0.371	2.11	<25	1.69
	294	0.746	2.53	42	<0.50

Although some of the permeate samples from Cell #1 and Cell #3 produced fluorescence readings in the range of the calibration curve, there was no conclusive evidence of microsphere passage. The scatter observed could be the result of extraneous particles being introduced into the collected permeate samples, such as corrosion byproducts in the permeate line, which artificially increased the fluorescence readings by beam reflection. Therefore, microsphere experiments were not conclusive for the three BW30HP coupons tested other than perhaps indicating that the coupons rejected more than 5 logs of microspheres. This is reinforced by the fact that microspheres were not expected to pass the BW30HP membrane due to previous work conducted by Adham *et al.* (1998a) which demonstrated complete MS2 *phage* rejection (>3.4 log – >7.4 log) by 11 replicates with BW30HP coupons.

ROGA - Solute Experiments

The concentration of feed solution constituents used for ROGA solute testing were 2,000 mg/L NaCl and 2,000 mg/L Na_2SO_4. Boric acid was eliminated as a testing solute due to its poor rejection by the ROGA membrane (0 to 15%). Figures 2.18–2.23 illustrate the data collected for three separate experimental sets (9/16/98, 9/23/98, and 10/5/98).

Figures 2.18–2.20 show 60 – 90% rejection of chloride ion over the entire range of flux values tested. Fitted *B* values between experimental sets seemed to correspond nicely with one another. For example, *B* values for the 9/23/98 experimental sets were slightly lower than those for the 9/16/98 experimental sets for all three cells. The largest *B* values for all three cells resulted from the 10/5/98 experimental set. There were some variations in *k* parameters in all three experimental sets. For Cells #2 and #3, the 10/5/98 experimental set experienced virtually no apparent concentration polarization (straight line nearly passing through the origin), however the same experimental set for Cell #1 experienced some concentration polarization. This is also apparent from the *k* parameter for Cell #2 and Cell #3 being much higher than the *k* parameter for Cell #1 for the 10/5/98 experimental set. The *k* parameters for Cells #2 and #3 were 1.072×10^5 m/d and 16.58 m/d respectively versus 1.990 m/d for Cell #1.

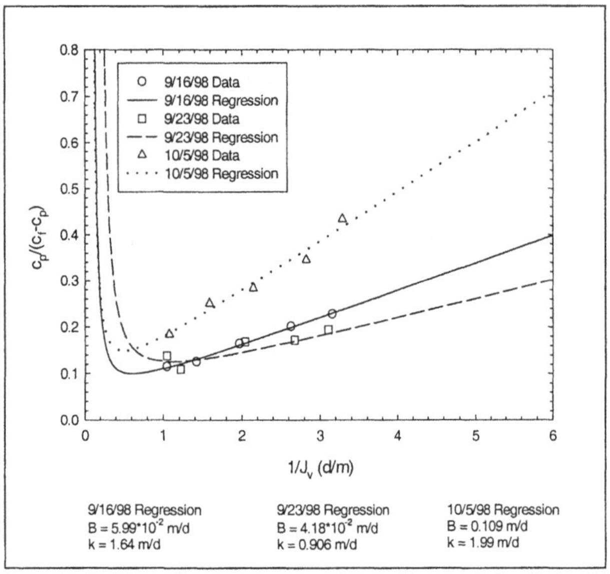

Figure 2.18 Experimental and model chloride ion passage for ROGA Cell #1

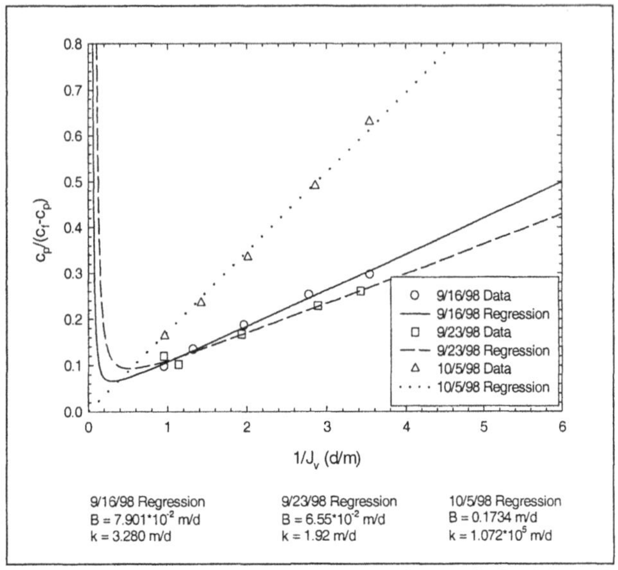

Figure 2.19 Experimental and model chloride ion passage for ROGA Cell #2

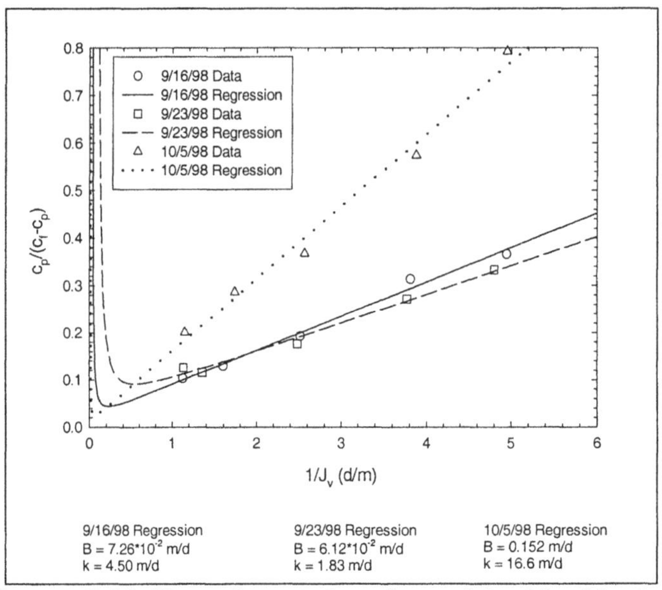

Figure 2.20 Experimental and model chloride ion passage for ROGA Cell #3

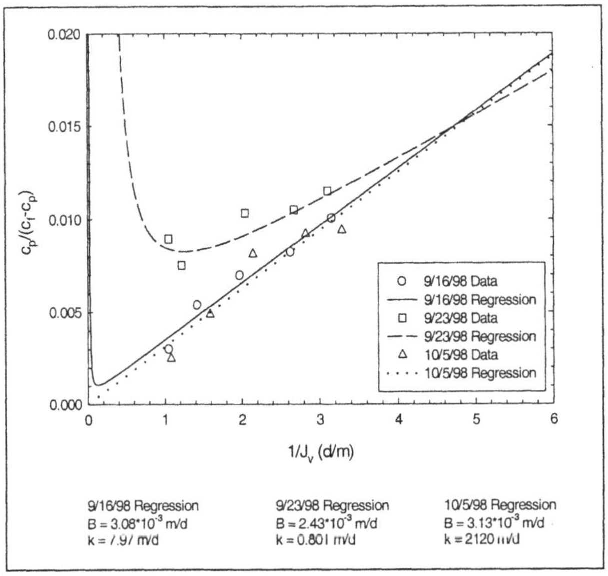

Figure 2.21 Experimental and model sulfate ion passage for ROGA Cell #1

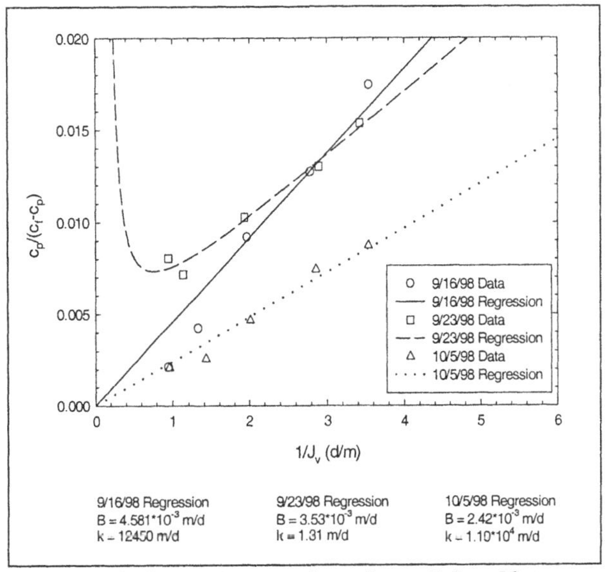

Figure 2.22 Experimental and model sulfate ion passage for ROGA Cell #2

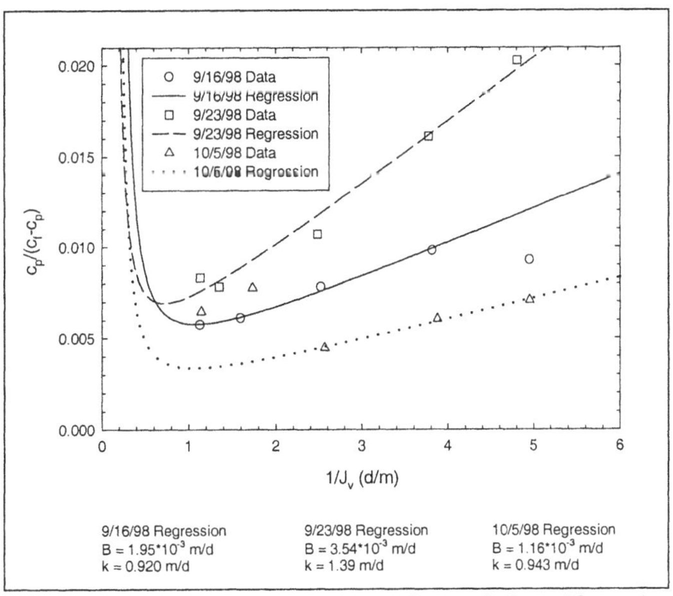

Figure 2.23 Experimental and model sulfate ion passage for ROGA Cell #3

This was most likely caused by slight system instability due to differences in corrosion products collected on the coupons. The data for all three experimental sets seemed to suggest that the system reached equilibrium with respect to the formation of corrosion products and the membranes experienced less fouling.

Figures 2.21–2.23 show sulfate ion rejections between 98 and 99.8%. The ROGA membranes seemed to be quite unstable with respect to sulfate ion rejection. In some instances, negative intercepts on the $c_p/(c_f - c_p)$ axis were observed and in other cases, relatively high levels of concentration polarization were observed. There is no apparent pattern in the changes in levels of concentration polarization with respect to levels of concentration polarization predicted by chloride ion rejection data.

ROGA - MS2 Phage Rejection

Two permeate and feed samples were analyzed for the presence of MS2 *phage* viruses. Although the assay procedure experienced a small amount of contamination, a seven log or greater rejection of MS2 *phage* was confirmed. MS2 *phage* rejection data for the ROGA coupons are summarized in Table 2.10.

MS2 *phage* rejection of greater than 7 log was in agreement with results reported by Adham *et al.* (1998a). Adham measured >4.5 - >6.0 log rejection of MS2 *phage* for a Fluid Systems ROGA membrane. The log rejection reported by Adham *et al.* (1998a) was limited by the concentration of MS2 *phage* in the feedwater since no MS2 *phage* was recovered in the permeate. Due to the complete rejection of MS2 *phage*, virus experiments with the ROGA coupons were discontinued, and microsphere testing was not pursued.

Table 2.10
MS2 *phage* virus rejection data for ROGA membranes

Cell	Pressure (psi)	Flux (m/d)	Virus Rejection
#1	451	0.90	> 7 log
#1	329	0.62	> 7 log
#2	419	1.00	> 7 log
#2	295	0.64	> 7 log
#3	399	0.87	> 7 log
#3	276	0.57	> 7 log

ESNA1 - Solute Experiments

The ESNA1 solute experiments were conducted using 2,000 mg/L NaCl and 2,000 mg/L Na_2SO_4. Solute and water permeation data for two experimental sets performed with the ESNA1 coupons on 12/17/98 and 12/29/98 (same membrane coupons for both sets) are shown in Figures 2.24–2.29 plotted according to Equation 1.6 as described previously. Inspection of these results revealed that chloride rejections ranged from 85 to 54% (Figures 2.24–2.26) and sulfate rejection were between 98.8 and 94.4% (Figures 2.27–2.29) over the flux range investigated. Another interesting characteristic displayed by the solute graphs was the dramatic increase in *k* values for all three cells between the data sets. This was not expected since an increase in *k* would indicate a decreased effect of concentration polarization on overall solute passage. From the graphs, the solute set collected on 12/29/98 was fitted with a higher *k* value than the solute set collected on 12/17/98. Therefore, in order for less concentration polarization to occur, the concentration polarization layer must decrease in thickness. This is not expected because the system is constantly subjected to corrosion by-products, which should deposit on the surface of the membrane, thereby, causing an increase in the thickness of the concentration polarization layer. As a general conclusion, several inconsistencies were observed between solute sets for both solutes investigated. The solute data is included in order to provide completeness instead of attempting to reveal the transport mechanism.

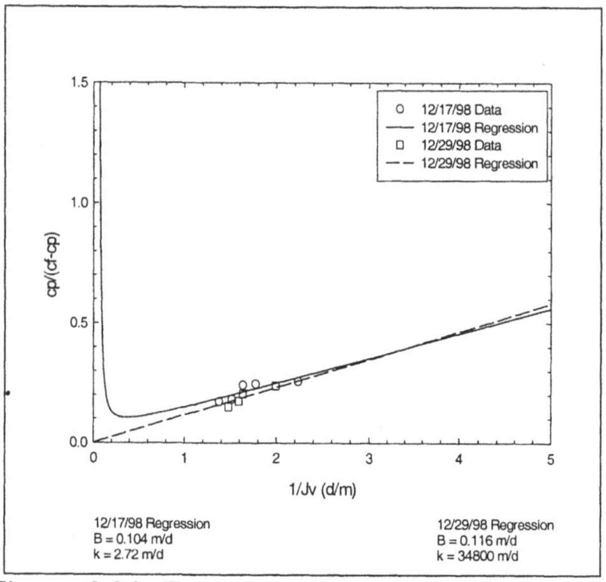

Figure 2.24 Experimental and model chloride ion passage for ESNA1 Cell #1

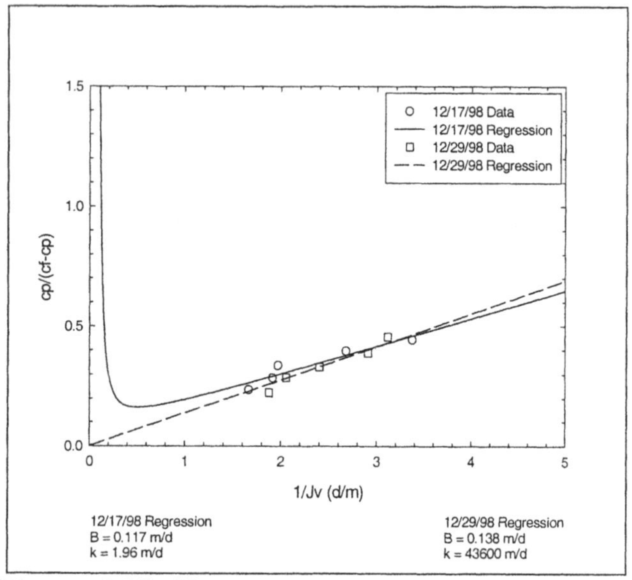

Figure 2.25 Experimental and model chloride ion passage for ESNA1 Cell #2

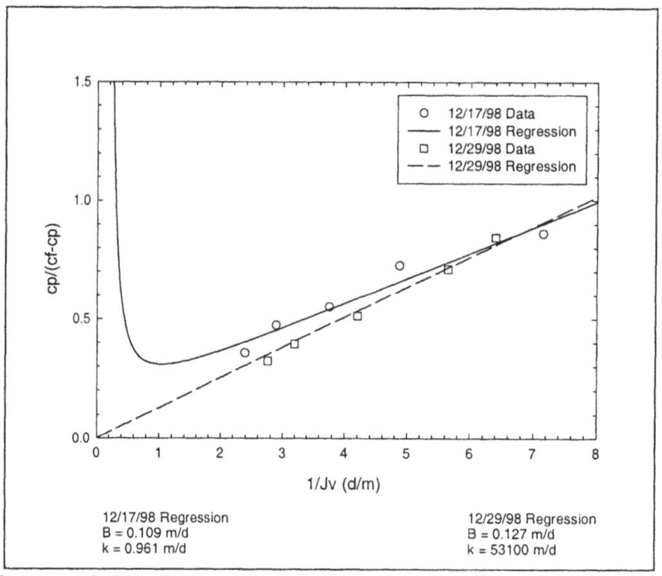

Figure 2.26 Experimental and model chloride ion passage for ESNA1 Cell #3

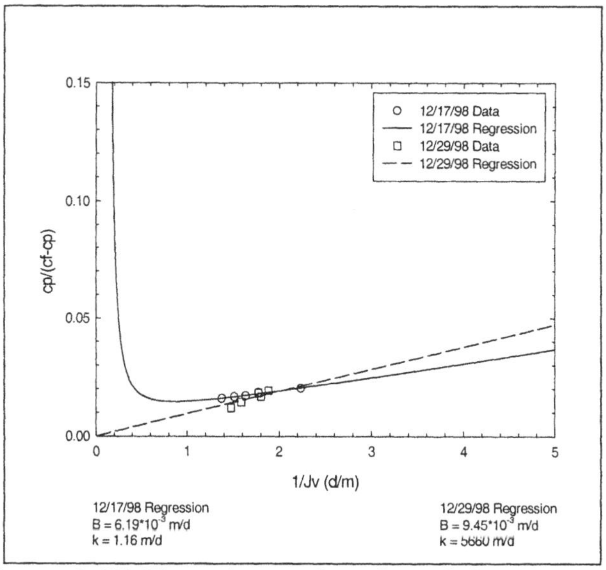

Figure 2.27 Experimental and model sulfate ion passage for ESNA1 Cell #1

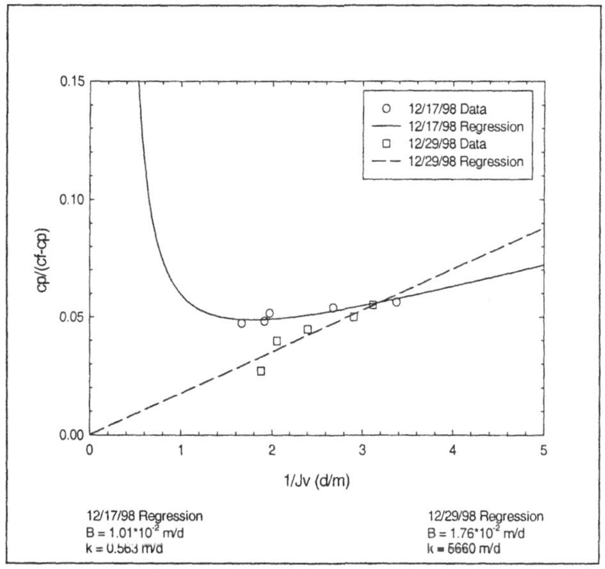

Figure 2.28 Experimental and model sulfate ion passage for ESNA1 Cell #2

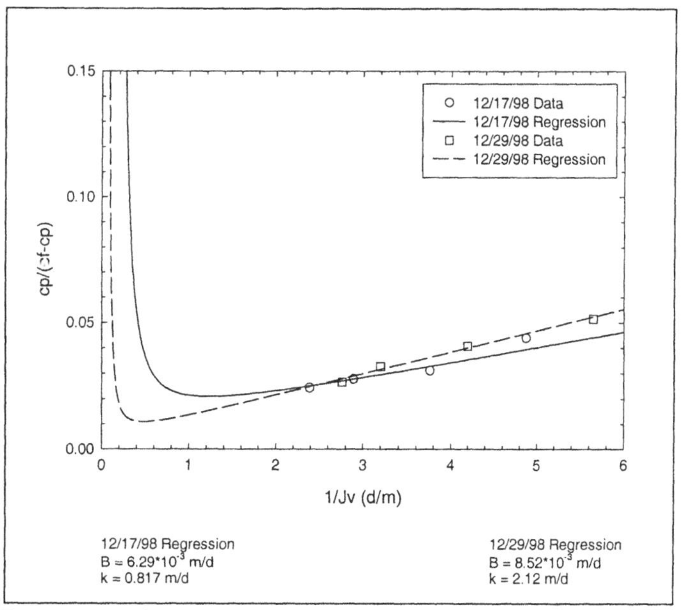

Figure 2.29 Experimental and model sulfate ion passage for ESNA1 Cell #3

ESNA1 - MS2 Phage Experiments

Due to the inability to assess the imperfections within the membrane with solute data, MS2 *phage* virus testing became the focus of the project in hopes that the higher sensitivity provided by virus data would uncover the presence of imperfections. Two separate MS2 *phage* tests were performed. The first experiment was performed on 12/29/98 using the same coupons used during solute testing. The second experiment was performed on 1/9/99 with newly installed coupons. The results of the MS2 *phage* experiments are discussed below.

Figure 2.30 summarizes the results for the 12/29/98 MS2 *phage* challenge test. Virus testing took place after the ESNA1 coupons were subjected to 14 days of continuous operation. From the chart, feed concentrations were in the range of 10^7–10^8 pfu/ml and permeate concentrations were <1 pfu/ml for all 15 permeate samples collected. The five different pressures tested resulted in a flux range of 5–18 gfd, typical of that to which membranes are subjected to in the field.

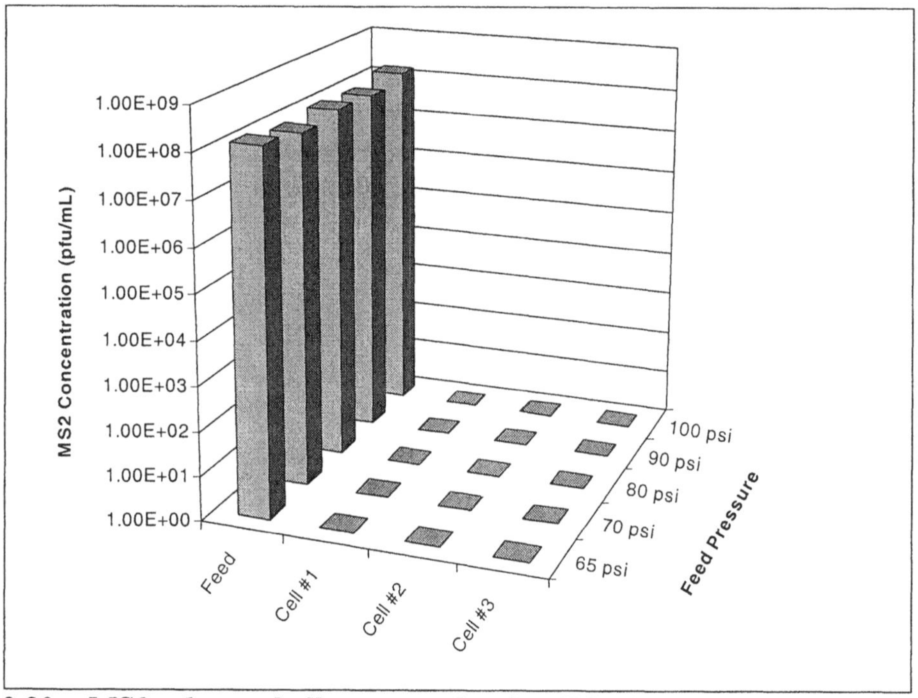

Figure 2.30 MS2 *phage* challenge results for ESNA1 coupons (12/29/98). Test performed 14 days post coupon installation.

Figure 2.31 illustrates the results for the 1/9/99 MS2 phage conducted 15 minutes after coupons were installed in the flat-leaf system. The bar chart shows feed concentrations between 10^7 and 10^8 pfu/ml and permeate concentrations of <1 pfu/ml. Therefore, ESNA1 coupons tested attained rejection of MS2 phage of > 7 – 8 logs.

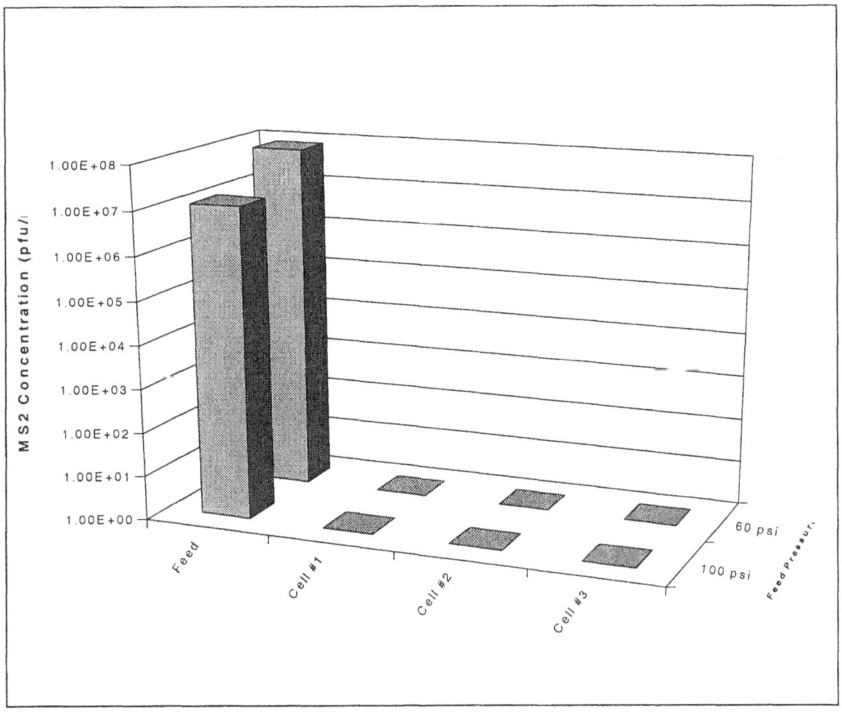

Figure 2.31 MS2 *phage* challenge results for ESNA1 coupons (1/9/99). Test performed 15 minutes post coupon installation.

CAB2 - Solute Experiments

The next membrane chosen to conduct solute and MS2 *phage* testing was the Hydranautics CAB2 membrane. A total of three spiral-wound elements were obtained from the full-scale plant of a drinking water utility. Two of the elements obtained were operated during startup procedures in the plant, then removed and shipped directly to the University of Illinois to be tested. One of the elements was never installed in the drinking water train, and, therefore, was considered to be the same as if obtained directly from the manufacturer.

The element previously installed in the first stage of the drinking water train was sacrificed, and coupons were cut and installed in the flat-leaf cells. Upon removing the fiberglass casing that housed the spiral-wound element, a thick layer of a white powdery substance was observed. Also, a relatively significant impression of the vexor was noted on the surface of the membrane.

Once the coupons were cut from the spiral-wound elements, they were installed in the flat-leaf cells and a solution containing only NaCl and Na_2SO_4 was allowed to flow over the surface under relatively low-pressure. This step was taken in an attempt to remove the white powder from the surface of the membrane before solute and MS2 *phage* testing began. As the system was brought to pressure, relatively large pressure fluctuations were observed due to the white powder obstructing flow through the needle valve being used to control system pressure. Virus testing did not begin until a steady, consistent pressure was observed indicating the white powder had been dissolved in the feed solution.

In addition to testing the coupons cut from the element that were installed in the drinking water utility, coupons cut from the brand new element were also tested. However, due to a laboratory mishap, only one sample was collected for both chloride and MS2 *phage* analysis from the brand new coupons.

The CAB2 solute experiments were conducted only to investigate whether the membranes were performing according to manufacturer's specifications. In addition, sulfate samples were diluted to a concentration below the typical calibration range for the IC analysis and their concentration could not be determined. As a result, chloride data was the only solute data collected successfully.

Two separate sets of solute and MS2 *phage* samples were collected for the coupons previously installed in the drinking water utility. One experiment was performed 15 minutes after installation (2/8/99) and consisted of two samples collected only for MS2 *phage* analysis. The second experiment was performed 10 days post installation (2/18/99) and consisted of an extensive solute test where five sets of solute samples were collected over a wide range of flux values. Only the set of solute samples collected on 2/18/99 was analyzed for chloride concentration. Figures 2.32–2.34 contain plots of the chloride solute data collected 10 days after installation in conjunction with MS2 *phage* samples from the element previously installed in the drinking water utility.

Two chloride samples collected from Cells #1 and #2 (Figures 2.32 and 2.33) contained much higher chloride concentrations than would be expected. Therefore, these points were omitted when regressing the data to obtain the fitted curve. Chloride rejections ranged from a maximum 96.8% rejection from Cell #1 to a minimum 89.2% rejection from Cell #3. Cells #2 and #3 both achieved a maximum 95% rejection of chloride at the highest flux rate tested. Manufacturer specifications provided with the element indicate a minimum chloride concentration of 97% after 30 minutes of operation. Therefore, chloride rejections of 95–96.8% were found to be satisfactory (i.e., no major imperfection resulting in gross solute leakage). The consistency with performance specifications was consistent with the dissolution of the relatively thick white powdery substance initially present on the membrane surface. The performance of the membrane should had deteriorated, if this fouling layer would have remained in place as a result of the correspondingly greater effect of concentration polarization.

CAB2 - MS2 Phage Experiments

The first MS2 *phage* experiment with the CAB2 coupons from the used element was conducted on 2/8/99. Samples were collected approximately 15 minutes after coupon set installation and addition of MS2 *phage*. Figure 2.35 is a plot showing the feed and the corresponding permeate concentrations of MS2 *phage* for both sets of collected samples. Approximately 10^8 pfu/ml were seeded in the feed tank and between 10^4 and 10^6 pfu/ml were found in the permeate samples over a flux range of 5–15 gfd. The CAB2 coupons that were obtained from an element previously installed in the full-scale plant of the drinking water utility only provided between 2 and 4 log rejection of MS2 *phage* after 15 minutes of operation in the flat-leaf system.

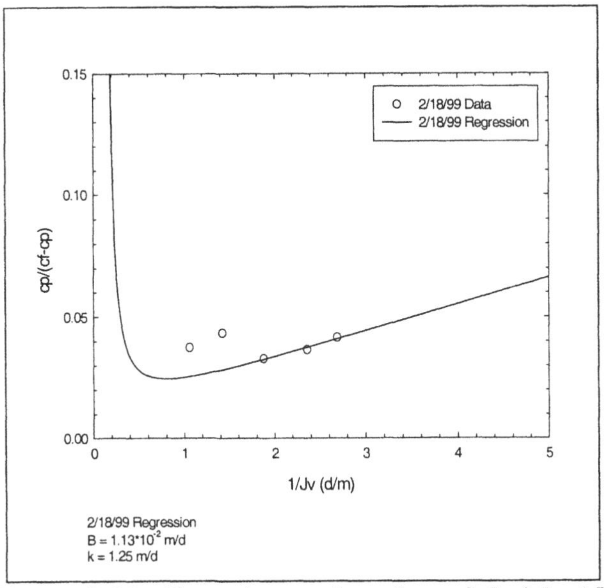

Figure 2.32 Experimental and model chloride ion passage for CAB2 Cell #1 (10 days post installation) (Coupons from used spiral-wound element)

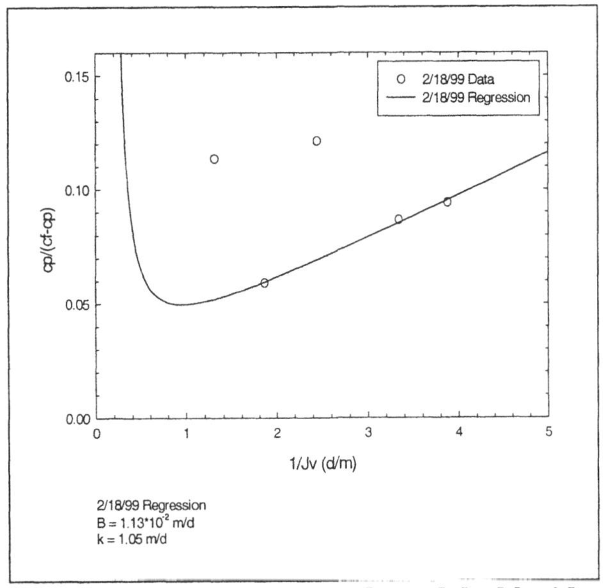

Figure 2.33 Experimental and model chloride ion passage for CAB2 Cell #2 (10 days post installation) (Coupons from used spiral-wound element)

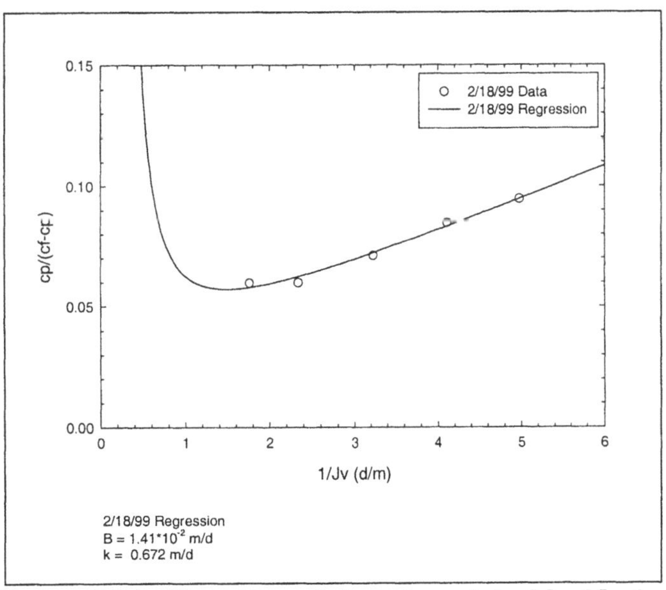

Figure 2.34 Experimental and model chloride ion passage for CAB2 Cell #3 (10 days post installation) (Coupons from used spiral wound element)

Once MS2 phage was detected in the 2/8/99 permeate samples, a more thorough experiment was conducted where a total of 15 samples were collected at 5 different pressure settings in order to investigate the role of flux in the passage of MS2 phage. This experiment was performed on 2/18/99, 10 days after the initial experiment, as mentioned previously. The MS2 phage concentrations for feed and permeate samples collected during the experiment are presented in Figure 2.36. Between 10^6 and 10^7 pfu/ml were injected into the feed solution, however, no MS2 phage was found in any of the permeate samples over a flux range of 5–28 gfd. The same coupons that previously only achieved 2–4 log MS2 phage rejection were now achieving greater than 6 log rejection. The next step was to replace the coupons being tested with coupons taken from the brand new element.

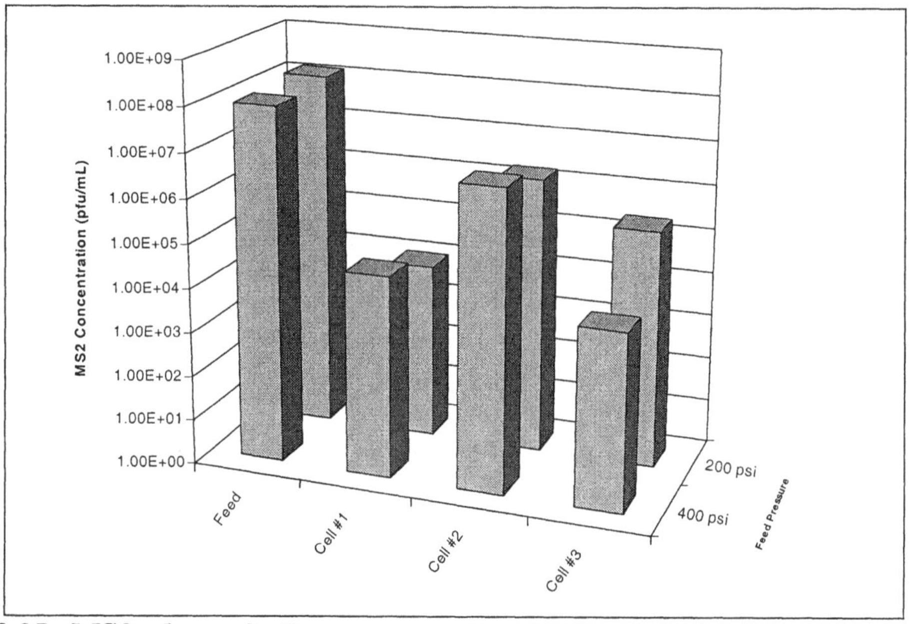

Figure 2.35 MS2 *phage* challenge results for CAB2 coupons cut from an element previously installed in the full-scale plant of a drinking water utility (15 minutes post installation)

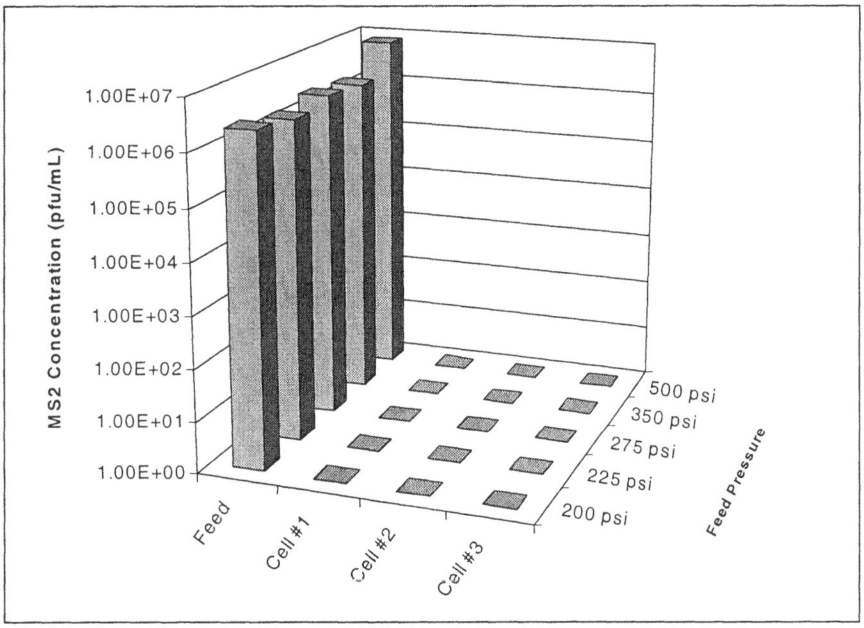

Figure 2.36 MS2 *phage* challenge results for CAB2 coupons cut from an element previously installed in the full-scale plant of a drinking water utility (10 days post installation)

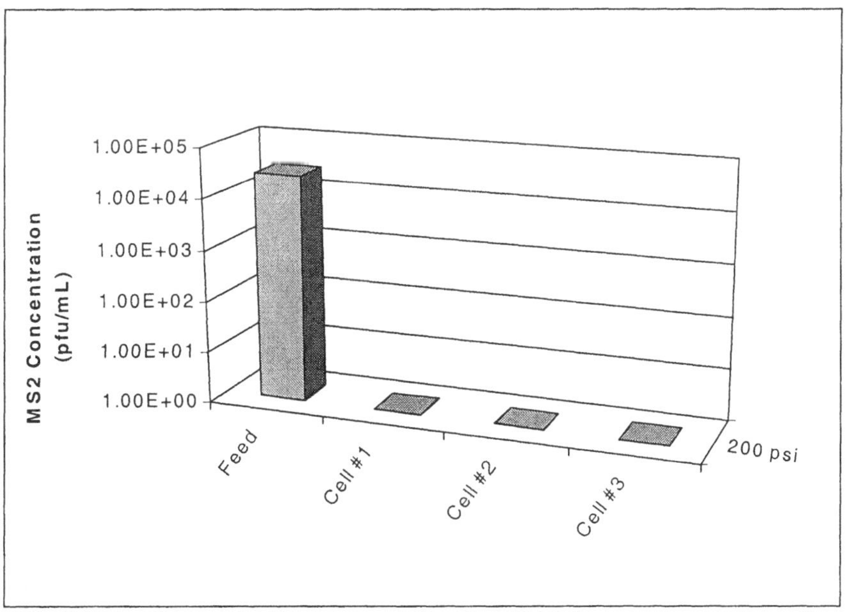

Figure 2.37 MS2 *phage* challenge results for CAB2 coupons taken from a new element (15 minutes post installation)

Tests with the coupons cut from the brand new element were performed on 2/22/99. MS2 *phage* testing was conducted approximately 15 minutes after the coupons were installed in the system. The results of the experiments are plotted in Figure 2.37. A feed MS2 *phage* concentration of 2.7×10^4 pfu/ml was accidentally used, which was significantly lower than previous MS2 *phage* feed concentrations. This was probably the result of a weak MS2 *phage* stock that was inadvertently used to inoculate the feed solution. Nevertheless, the coupons achieved total MS2 *phage* rejection which amounted to >4.4 log rejection over a flux range of 7-9 gfd. Therefore, the new coupons were achieving somewhat greater MS2 *phage* rejection than the coupons taken from the used element. However, these differences might be the result of variability in membrane performance from element to element or from coupon to coupon within an element.

ESPA1 - Solute Experiments

Once the time dependence of MS2 *phage* rejection was discovered with the Hydranautics CAB2 membrane coupons, Hydranautics ESPA1 coupons were installed in order to investigate reproducibility of this phenomenon with a different membrane. Experiments were performed with coupons cut from flat-leaf sheets obtained from the manufacturer.

The initial experiment with ESPA1 coupons was performed on 3/15/99 and involved collecting two separate samples at different pressure settings immediately after coupon installation. These samples were tested for chloride, sulfate, and MS2 *phage* concentrations. After the initial experiment, another experiment was performed 24 hours later with the same coupons on 3/16/99 at 5 different flux rates.

The two sets of samples collected on 3/15/99 were analyzed for both chloride and sulfate concentrations. Since only two samples were collected, the data was not plotted according to Equation 1.6. This set of solute data was used solely for the purpose of determining if a major imperfection was accidentally induced during coupon installation (i.e., a hole in the membrane). The solute rejections for the two samples collected on 3/15/99 are presented in Table 2.11.

The results from the second set of collected solute samples (3/16/99) are plotted in Figures 2.38–2.43. Solute results from the 3/16/99 experimental set show the possibility that membrane imperfections may have played a key role in the overall solute passage for the Hydranautics ESPA1 membrane. From initial inspection of Figures 2.38–2.43, the regressions performed on each solute set independently reveal k values for sulfate much lower than those for chloride (see Table 2.12). Expected k values for chloride were approximately 1.4 times those for sulfate based solely on diffusivity values for chloride of 1.49×10^{-5} cm^2/s and sulfate of 1.05×10^{-5} cm^2/s. However, the actual ratio of k values shown in Table 2.12 were 2.81 for Cell #1, 2.64 for Cell #2, and 4.22 for Cell #3, indicating that the assumption of $\varepsilon = 0$ may not be valid. In addition, the regression curves shown in Figures 2.41–2.43 do not represent the observed data pattern. In order to explain the relationship between the chloride and sulfate solute data, Equation 1.8, which takes into account passage through imperfections, was chosen as an alternative to Equation 1.6.

Table 2.11
Chloride and sulfate rejections for 3/15/99 solute samples

Cell	Pressure (psi)	Flux (m/d)	Chloride Rejection (%)	Sulfate Rejection (%)
1	99	0.617	96.8	99.32
1	203	1.620	98.6	99.73
2	82	0.444	96.1	99.52
2	186	1.540	98.4	99.82
3	62	0.196	92.3	98.90
3	167	1.176	98.0	99.87

Table 2.12
Summary of *k* values for 3/16/99 data set assuming ε = 0

Cell	k_{NaCl} (m/d)	k_{Na2SO4} (m/d)	k_{NaCl} / k_{Na2SO4}
1	3.29	1.17	2.81
2	2.45	0.928	2.64
3	2.76	0.654	4.22

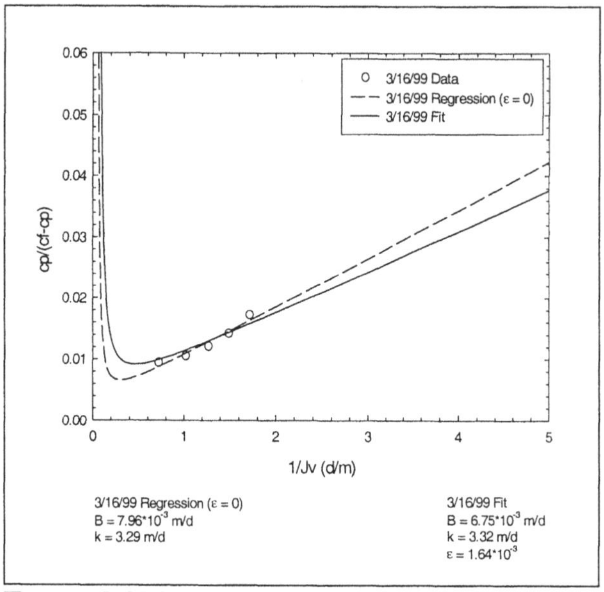

Figure 2.38 Experimental and model chloride ion passage for ESPA1 Cell #1

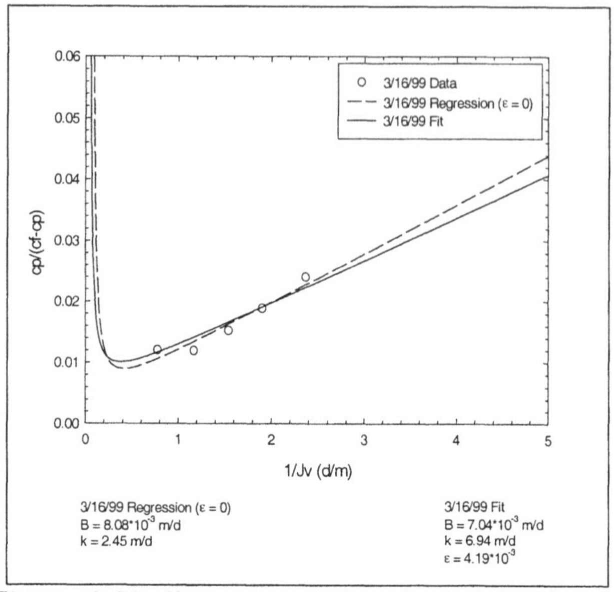

Figure 2.39 Experimental and model chloride ion passage for ESPA1 Cell #2

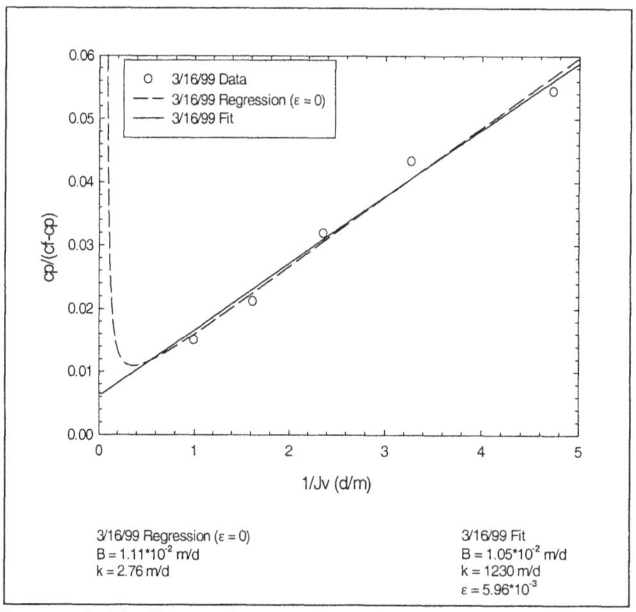

Figure 2.40 Experimental and model chloride ion passage for ESPA1 Cell #3

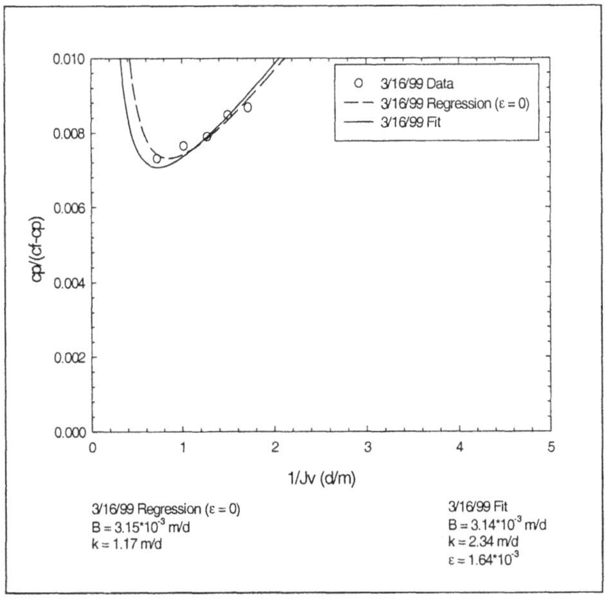

Figure 2.41 Experimental and model sulfate ion passage for ESPA1 Cell #1

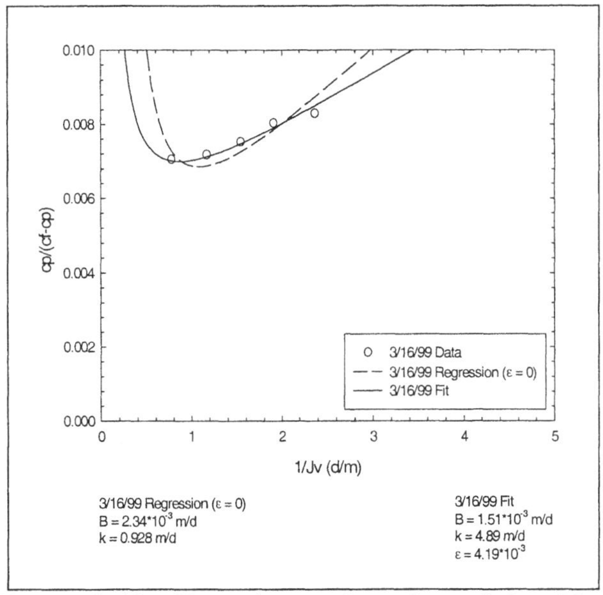

Figure 2.42 Experimental and model sulfate ion passage for ESPA1 Cell #2

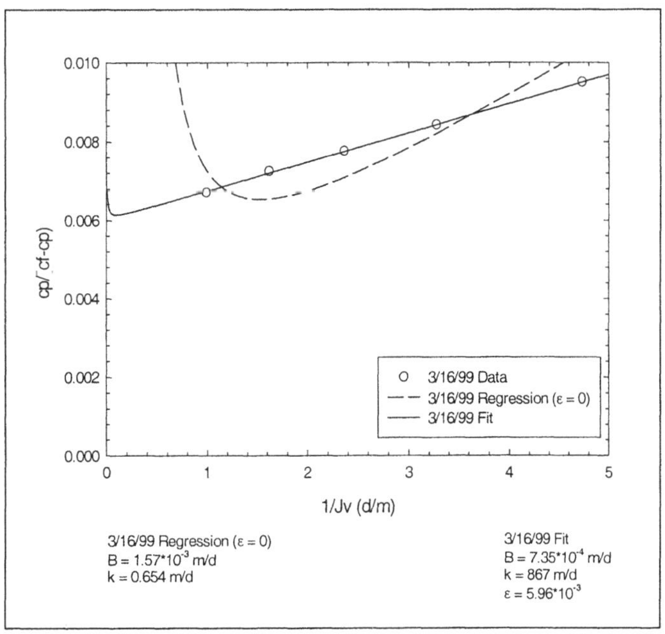

Figure 2.43 Experimental and model sulfate ion passage for ESPA1 Cell #3

Table 2.13
Summary of fitting parameters obtained with Equations 1.6 and 1.8

Cell	Equation 1.6				Equation 1.8				
	k_{NaCl} (m/d)	B_{NaCl} (m/d)	k_{Na2SO4} (m/d)	B_{Na2SO4} (m/d)	k_{NaCl} (m/d)	B_{NaCl} (m/d)	k_{Na2SO4} (m/d)	B_{Na2SO4} (m/d)	ε
1	3.29	7.96×10^{-3}	1.17	3.15×10^{-3}	3.32	6.75×10^{-3}	2.34	3.14×10^{-3}	1.64×10^{-3}
2	2.45	8.08×10^{-3}	0.928	2.34×10^{-3}	6.94	7.04×10^{-3}	4.89	1.51×10^{-3}	4.19×10^{-3}
3	2.76	1.11×10^{-2}	0.654	1.57×10^{-3}	1230	1.05×10^{-2}	867	7.35×10^{-4}	5.96×10^{-3}

In order to obtain values for ε, k, and B, a nonlinear least squares fit was performed with the ratio of k values between chloride and sulfate fixed based on diffusivity values given previously. The regressions were performed for the chloride and sulfate data together in a unified regression, thereby finding k values that best predict observed solute data while maintaining the proper ratio. The resulting fits are labeled as "Fit" and plotted in Figures 2.38–2.43. Parameter values for Equations 1.6 and 1.8 are summarized in Table 2.13. Comparing the fitting results from both equations, Equation 1.8 resulted in a much better fit. The solute leakage ε predicted by fitting the data with Equation 1.8 ranged from 1.64x10-3 for Cell #1 to 5.96x10-3 for Cell #3.

ESPA1 - MS2 Phage Experiments

As outlined previously, two separate virus experiments were performed with the ESPA coupons. The first experiment took place on 3/15/99 approximately 15 minutes after the coupons were installed in the flat-leaf system. The results of the 3/15/99 MS2 *phage* experiment are shown in Figure 2.48. The results for the second experiment performed on 3/16/99 approximately 24 hours after coupon installation are summarized in Figure 2.45.

During the experiment conducted on 3/15/99, approximately 10^8 pfu/ml of MS2 *phage* were seeded in the feed tank. Rejection levels ranged from 5 logs for Cell #2 to 7 logs for Cells #1 and #3, or approximately 2 to 4 logs less passage of viruses compared to those estimated from solute data (see ε in Table 2.13). Consequently, the assumption that leakage of solute is in direct relationship with passage of viruses may be somewhat invalid. The imperfections that allow solute leakage across the membrane may not be large enough to allow MS2 *phage* passage.

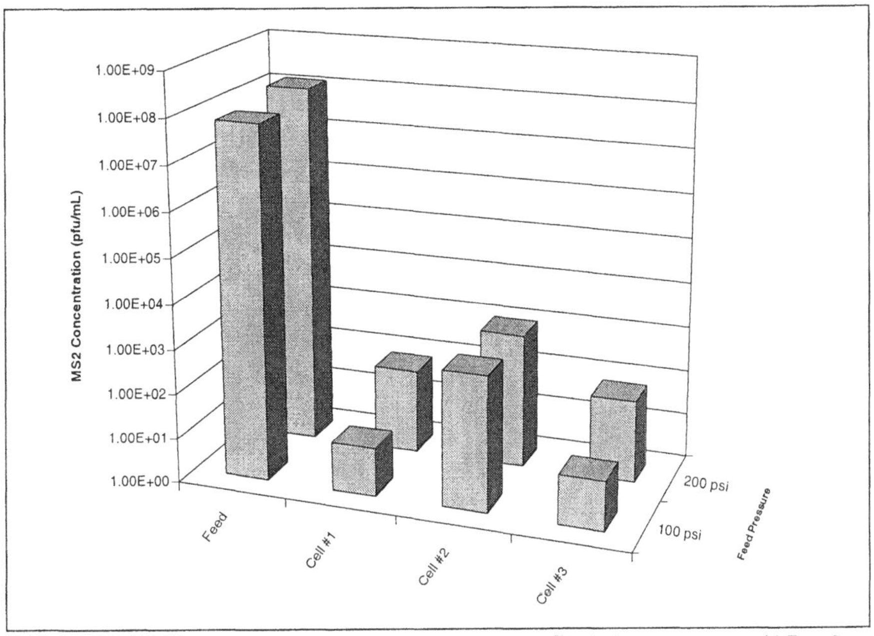

Figure 2.44 MS2 *phage* challenge results for ESPA1 coupons (15 minutes post installation)

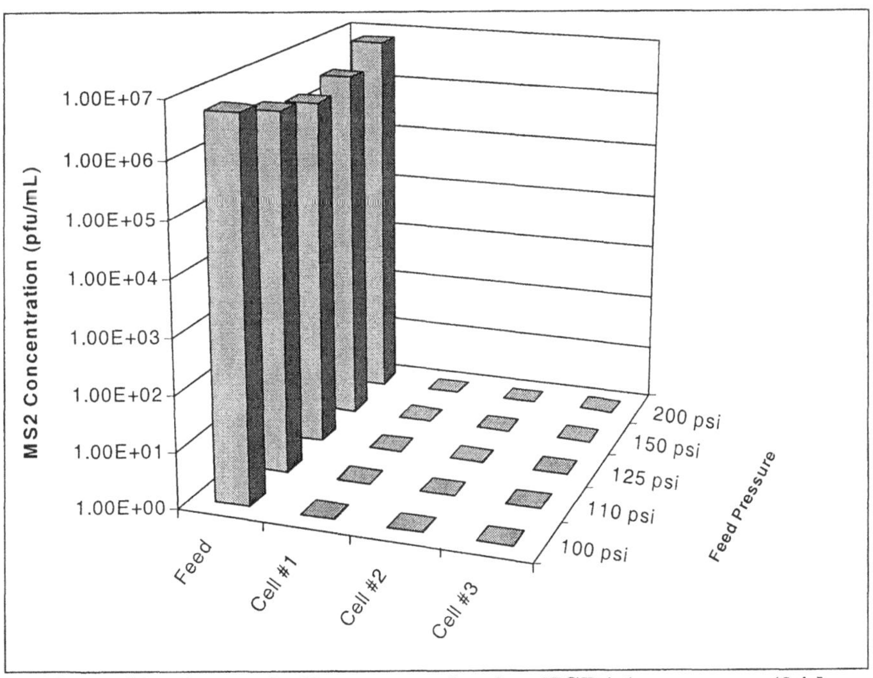

Figure 2.45 MS2 *phage* challenge results for ESPA1 coupons (24 hours post installation)

For all three of the ESPA1 coupons tested, the ε value was in the order of 10^{-3}. Therefore, if this level of leakage was also allowing MS2 *phage* passage, only 3 log rejection of MS2 *phage* would be expected if the size of the imperfections are ignored and the assumption that an imperfection that allows leakage of solute will also allow passage of MS2 *phage* is valid. Obviously, this was not the case due to the increased levels of MS2 *phage* rejection that were observed. Therefore, the occurrence of solute leakage (i.e. ε >0) should only be used as an indicator for the possibility of MS2 *phage* passage.

The analysis of samples collected from the experiment performed on 3/16/99, 24 hours after coupon installation, revealed all coupons achieving complete MS2 *phage* rejection (Figure 2.45). A concentration of 10^6 to 10^7 pfu/ml in the feed solution produced no MS2 *phage* in the permeate. Therefore, the ESPA1 coupons were achieving >6 log rejection. The time dependence of MS2 *phage* passage observed initially with the CAB2 coupons was reproduced with ESPA1 coupons. Apparently, clogging of the imperfections was also taking place.

ESPA1 - Air Entrainment Experiments

Air entrainment tests were conducted on 3/26/99 with ESPA1 coupons previously installed on 3/19/99. The coupons used had previously achieved complete MS2 *phage* rejection during an experiment conducted on 3/19/99. The first set of samples on 3/26/99 were taken before the addition of MS2 *phage* to the feed tank to show no virus contamination in the permeate lines. As a result, no MS2 *phage* was recovered in these permeate samples. The second sample was taken after exposing the membranes for approximately 10 minutes to air bubbles in the feedwater. Experimental results are shown in Figure 2.46.

An additional set of samples was collected approximately 24 hours after air entrainment was discontinued. These samples were taken to investigate whether permanent damage was inflicted on the ESPA1 coupons by the air entrainment. The results are also shown in Figure 2.46. Figure 2.46 shows the absence of MS2 *phage* in the feed and permeate before seeding and air entrainment. However, once MS2 *phage* was seeded in the feed tank and air bubbles were introduced in the feedwater, MS2 *phage* was detected in the permeate samples for Cells #1 and #2 (During Air). Approximately 10^8 pfu/ml of MS2 *phage* was added to the feed tank resulting in log rejections of 7.3 for Cell #1, 6.9 for Cell #2, and >8 for Cell #3. After the system was allowed to stabilize under the absence of air from the feedwater, MS2 *phage* rejections for all three cells were >6.8 log. However, due to the lower MS2 *phage* feed concentration (6.0×10^6 pfu/ml), a change in the MS2 *phage* rejection during and after the air entrainment could not be assessed. Nevertheless, MS2 *phage* rejections on the order of 7 log can be considered excellent for drinking water treatment purposes.

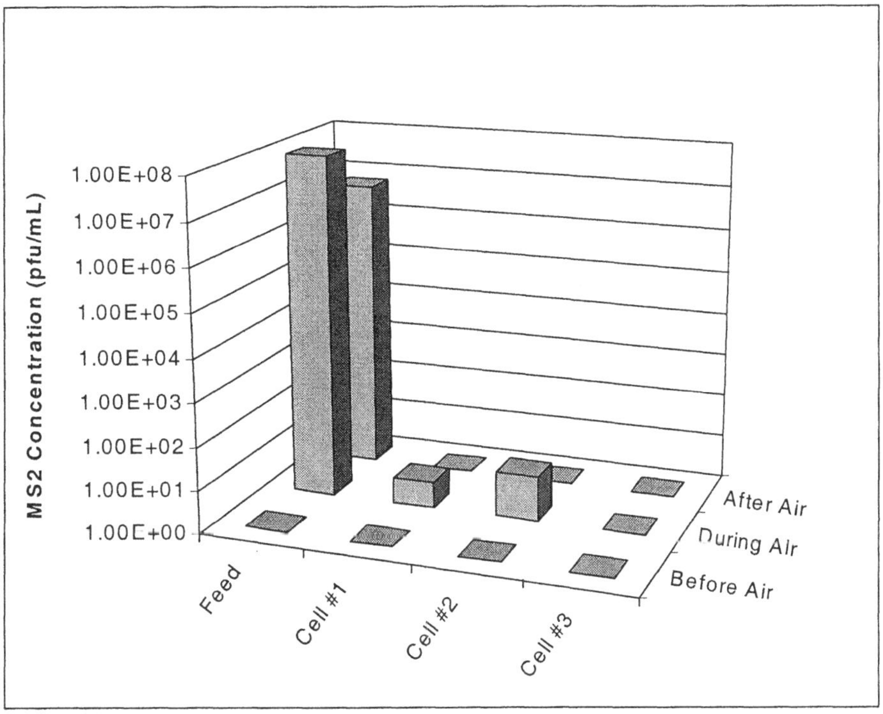

Figure 2.46 MS2 *phage* challenge results for air entrainment experiments with ESPA1 coupons

ESPA1 - Induced Imperfection Experiments

Imperfection Characterization. The relatively high levels of MS2 *phage* rejection exhibited by membranes tested including the ESPA1 could not be correlated to similar rejection levels for microspheres due to analytical detection limitations for the non-biological surrogate indicators. Consequently, physical imperfections were induced as described previously in order to correlate microsphere rejection to virus removal. The induced imperfections were characterized after testing by obtaining digital photographs at 10× magnification using light microscopy (Figures 2.47-2.52). The imperfections were estimated to be approximately 150 μm in diameter. However, the severity of the imperfection differed between the three coupons. Cell #2 contained the most severe imperfection as this imperfection allowed the most light to pass when viewed under the microscope. Cells #1 and #3 seemed to contain similar imperfections from inspection of the micrographs. In addition, the micrographs for Cell #3 showed the presence of a rather large particle obstructing a portion of the imperfection. This particle may explain the dynamic nature of the MS2 *phage* rejection observed for Cell #3.

ESPA1 MS2 Phage Experiments with Imperfection. Initially, three ESPA1 coupons with an induced "scratch" were tested for virus rejection on 5/10/99. However, even though the active layer was scratched with a sewing needle, all three cells achieved complete virus rejection (>6.3 log) apparently because the polysulfone layer was not ruptured. Therefore, three additional coupons were installed and tested on 5/14/99 in order to confirm this observation. When inducing the imperfections for Cells #1 and #3, a rather large portion of the active layer was removed. Again, all three cells achieved complete rejection (>5.7 log). Therefore, the polysulfone layer, essentially an ultrafiltration membrane, was exhibiting high levels of MS2 *phage* rejection.

Additional virus experiments with the ESPA1 coupons were conducted on 5/19/99 to 5/25/99. Three new ESPA1 coupons were installed containing imperfections consisting of a hole approximately 150 μm in diameter as previously described and characterized. Results from these virus challenge tests are shown in Figures 2.53-2.55 and summarized in Table 2.14.

Table 2.14
Summary of MS2 *phage* experiments for ESPA1 coupons with imperfections

Date	Cell	Pressure (psi)	Flux (m/d)	MS2 Rejection (log)
5/19/99	1	103	0.626	>4.7
	1	198	1.50	>5.0
	2	91	2.49	1.3
	2	186	4.59	1.3
	3	79	0.566	>4.7
	3	173	1.29	>5.0
5/21/99	1	198	1.25	>4.7
	1	100	0.510	>5.0
	2	186	3.74	0.31
	2	88	2.21	0.34
	3	173	1.03	3.1
	3	76	0.305	4.0
5/25/99	1	100	0.467	>8.6
	1	199	1.17	>8.6
	2	87	1.70	0.11
	2	187	2.96	0.46
	3	76	0.277	>8.6
	3	174	0.954	>8.6

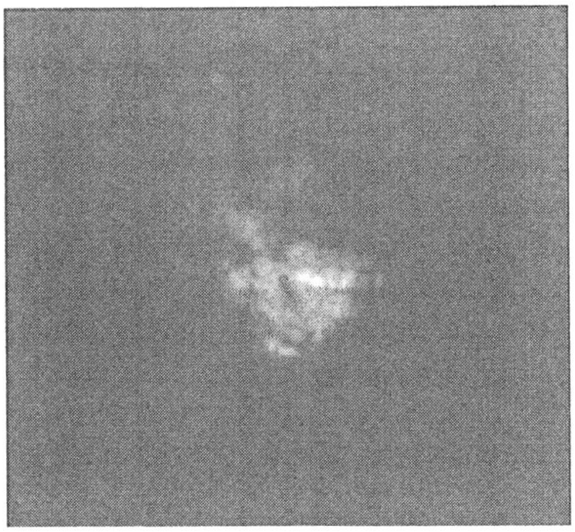

Figure 2.47 Micrograph of induced pinhole on Coupon #1 taken from the back

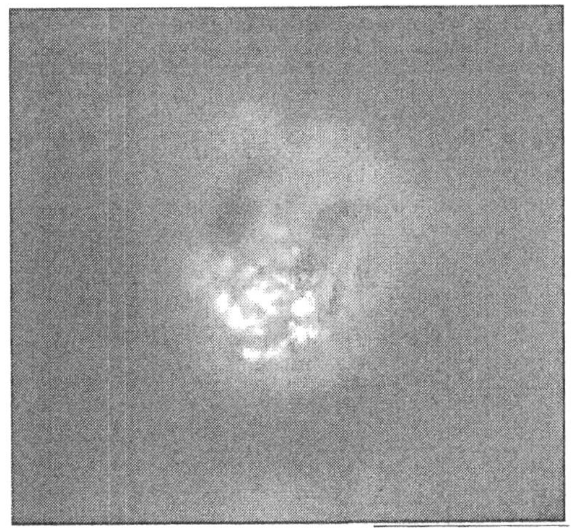

Figure 2.48 Micrograph of induced pinhole on Coupon #1 taken from the front

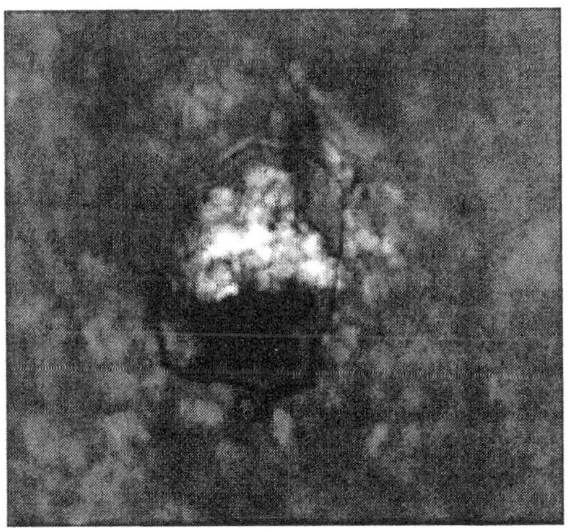

Figure 2.49 Micrograph of induced pinhole on Coupon #2 taken from the back

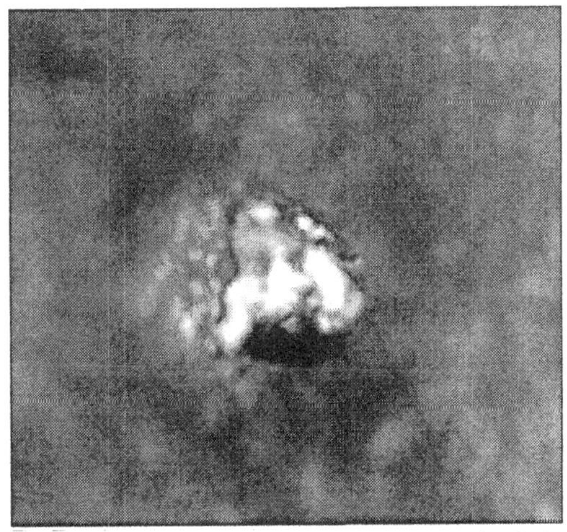

Figure 2.50 Micrograph of induced pinhole on Coupon #2 taken from the front

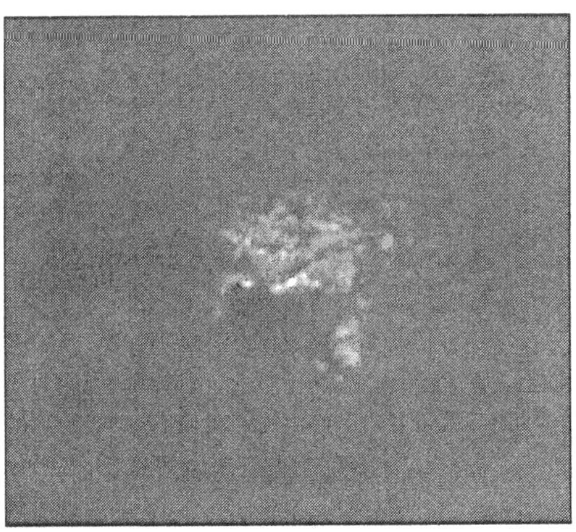

Figure 2.51 Micrograph of induced pinhole on Coupon #3 taken from the back

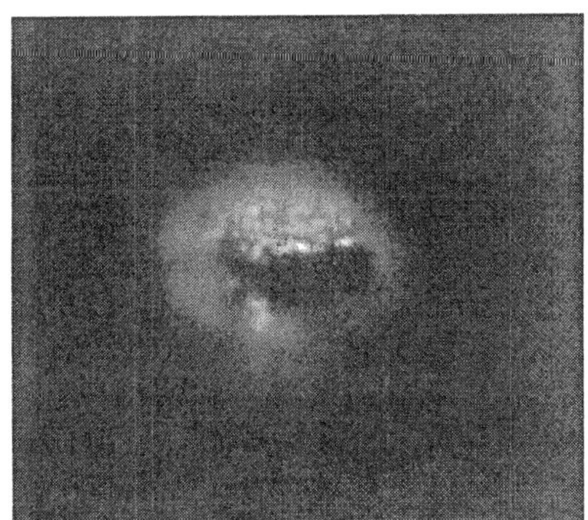

Figure 2.52 Micrograph of induced pinhole on Coupon #3 taken from the front

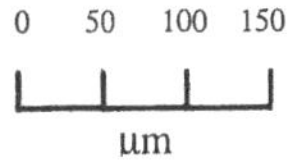

The imperfection induced in the membrane coupon of Cell #1 did not result in noticeable virus passage. MS2 *phage* rejection was >4.7–8.6 log as no *phage* was recovered in the permeate samples for any of the experiments conducted with this particular membrane coupon. Characterization of the imperfection after testing by light microscopy revealed that the imperfection might not have completely penetrated the polysulfone layer.

The membrane coupon in Cell #2 provided MS2 *phage* rejections ranging from 0.11 log to 1.3 log over the three separate experiments conducted. Consistent with the observations, the imperfection induced in this particular coupon was more severe than the other two (see Figures 2.47–2.52). The polysulfone layer was penetrated more deeply, thereby causing the dramatic increase in flux and increase in MS2 *phage* passage.

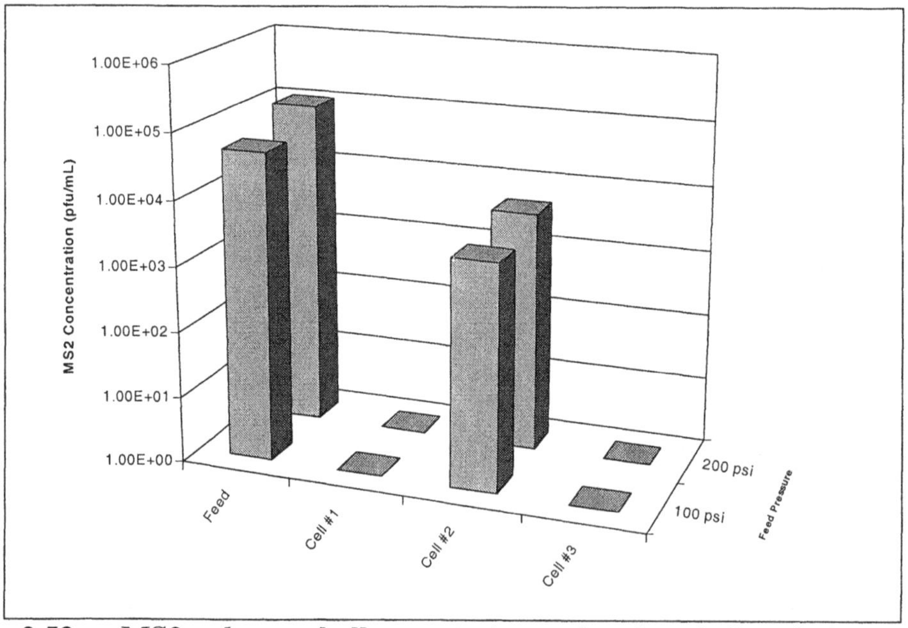

Figure 2.53 MS2 *phage* challenge test results for induced imperfection experiments with ESPA1 coupons (5/19/99)

Inspection of MS2 *phage* rejection data for Cell #3 revealed a more dynamic effect. Initially, the membrane coupon in Cell #3 gave complete virus rejection (Figure 2.53). Then, during the experiment performed on 5/21/99, MS2 *phage* viruses were recovered in the permeate from Cell #3, corresponding to rejection ranging from 3.1–4.0 log. However, during the final experiment, performed on 5/25/99, the cell again achieved rejections in excess of 8.6 logs. This may have been due in part to the rather large obstruction observed in the micrographs for Cell #3. The increase in virus passage between 5/19/99 and 5/21/99 might have been the result of a transient initial performance. Even though steady state with respect to solute is reached relatively quickly, longer operating times might be required for virus passage to stabilize. This observation for Cell #3 was consistent with the increase in virus rejection for Cell #2 from 5/19/99 to 5/21/99. Then, at some point in time between the sampling events on 5/21/99 and 5/25/99, a large particle obstructed the imperfection of Cell #3, thereby, allowing an increased level of MS2 *phage* rejection by decreasing the flow through the imperfection. The poor virus rejection provided by the coupon in Cell #2 was approximately the same on 5/21/99 and 5/25/99.

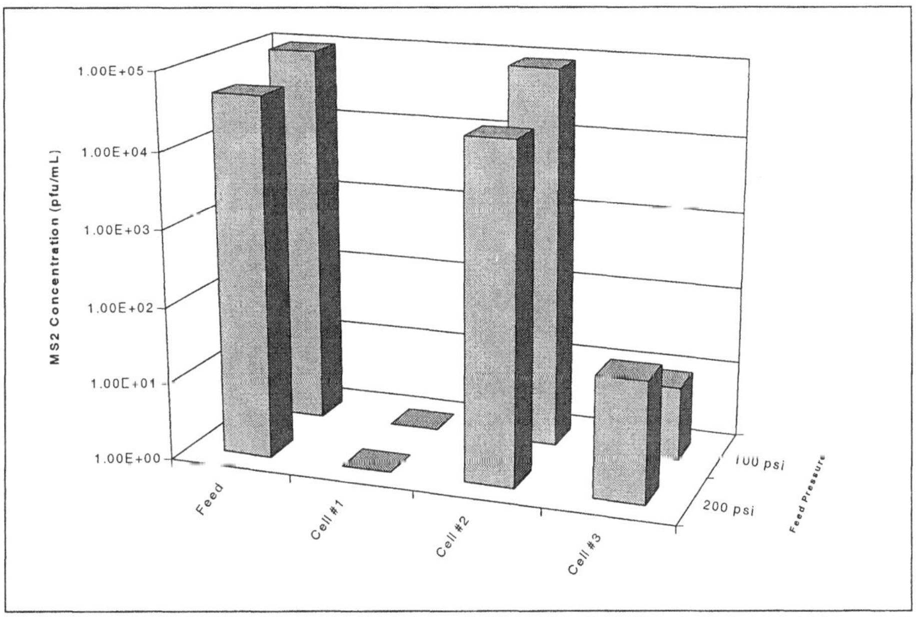

Figure 2.54 MS2 *phage* challenge test results for induced imperfection experiments with ESPA1 coupons (5/21/99)

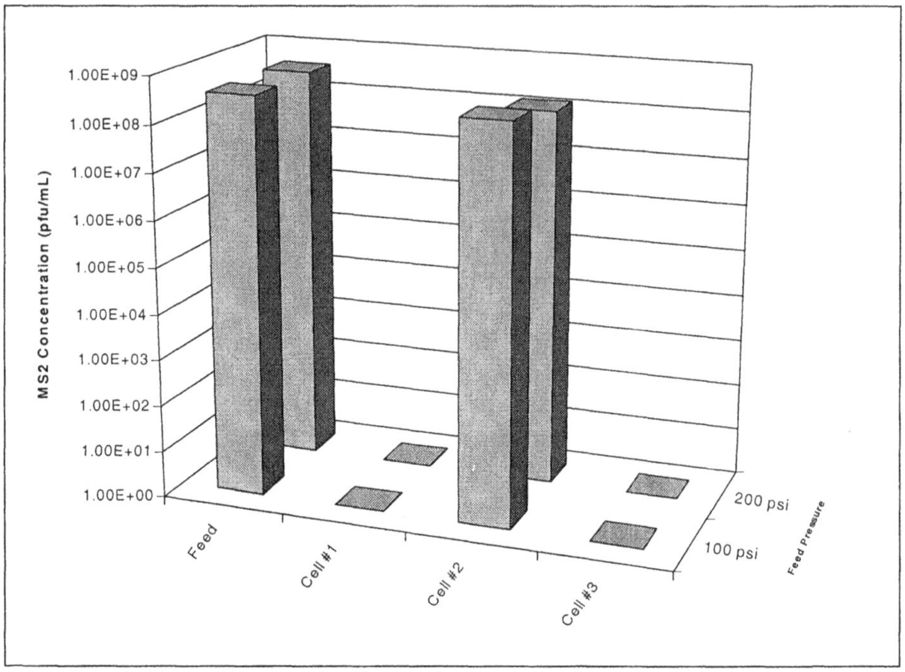

Figure 2.55 MS2 *phage* challenge test results for induced imperfection experiments with ESPA1 coupons (5/25/99)

ESPA1 Microsphere Experiments with Imperfection

Microsphere experiments were also conducted using the same ESPA1 coupons with induced imperfections described in the preceding section. Microsphere experiments performed on 5/21/99 took place directly after MS2 *phage* experiments were performed. Microsphere experiments performed on 5/25/99 were conducted in conjunction with MS2 *phage* experiments as both virus and microsphere stock were added to the feed tank at the same time and collected samples were analyzed for both microsphere and MS2 *phage* concentrations.

Experimental results for the microsphere tests conducted on 5/21/99 and 5/25/99 are summarized in Table 2.15. Fluorescent readings for permeate samples collected from Cell #1 and #3 were at the low end within the range of the calibration curve used at which a high noise level was observed. Consequently, a specific concentration of microspheres was not calculated.

ESPA1 MS2 phage and Microsphere Rejection Correlation

A correlation between MS2 *phage* and microsphere rejection could not be found due to detection limitation for the microspheres at the relatively high rejection levels observed for Cells #1 and #3. However, a good correlation was observed for Cell #2 which gave rejections of 0–0.5

log for both indicators. But perhaps this is the range of practical interest because a relatively major imperfection (0–0.5 log rejection) in a flat-leaf coupon would translate into a relatively minor imperfection (2–4 log rejection) in a single spiral-wound element due to the difference in surface area. Furthermore, the same level could correspond to nearly 5 log for vessels containing various elements with just one such imperfection in one element.

Table 2.16 summarizes the observed correlation between MS2 *phage* and microsphere rejection for Cell #2. For each of the first three comparisons, microsphere and MS2 *phage* rejections correlate 1:1 when taking into consideration analytical variability. The last comparison (5/25/99–2nd sample) is the only sample that did not directly correlate. At the time this sample was taken, a fairly large amount of fouling was occurring in the system due to high concentrations of microspheres and organics in the feed solution. This may explain the greater difference between the MS2 *phage* and microsphere rejections.

Table 2.15
Summary of microsphere experiments for ESPA1 coupons with imperfections

Date	Cell	Pressure (psi)	Flux (m/d)	Predicted Microsphere Conc.	Feed Microsphere Conc.	Microsphere Rejection (log)
5/21/99	1	100	0.503	<300 ng/L	73 μg/L	>2.4
	1	198	1.22	<300 ng/L	179 μg/L	>2.8
	2	88	2.13	35 μg/L	73 μg/L	0.32
	2	186	3.66	98 μg/L	179 μg/L	0.26
	3	76	0.304	<200 ng/L	73 μg/L	>2.6
	3	174	1.02	<400 ng/L	179 μg/L	>2.7
5/25/99	1	100	0.467	<200 ng/L	8.13 mg/L	>4.6
	1	199	1.17	<200 ng/L	7.39 mg/L	>4.6
	2	87	1.70	5.77 mg/L	8.13 mg/L	0.15
	2	187	2.96	5.29 mg/L	7.39 mg/L	0.15
	3	76	0.277	<200 ng/L	8.13 mg/L	>4.6
	3	174	0.954	<200 ng/L	7.39 mg/L	>4.6

Table 2.16
Correlation of MS2 *phage* and microsphere rejection for Cell #2

Date	Flux (m/d)	MS2 *phage* Rejection (log)	Microsphere Rejection (log)
5/21/99	2.1 – 2.2	0.34	0.32
	3.7	0.31	0.26
5/25/99	1.7	0.11	0.15
	3.0	0.46	0.15

Spiral-wound Element Experiments

BW30HP - Solute Experiments

Experimental results obtained for all five experiments performed with the BW30HP element are presented in Figure 2.56. As depicted in the plots for water and sodium chloride, good reproducibility was observed prior to chlorine treatment (i.e., Experiments 1-3). A comparison to manufacturer's performance specifications revealed that the permeate flow rate observed for Experiments 1-3 in this study was approximately half of that expected and the NaCl rejection was in excess of the average of 99 percent for a new element. These findings are not very surprising because the element used was installed in the unit for nearly 8 years after use in various other projects. Some level of membrane compaction would be expected for such an aged element. This element was tested in this study for the purpose of assessing if an older version of a membrane reported to serve as an excellent barrier for MS2 *phage* when new (Adham *et al.*, 1998a, 1998b) would allow virus passage. Virus testing results are discussed in the following section. Boric acid data were more variable due to analytical problems. Nevertheless, the rejection of this neutral weaker electrolyte molecule was lower compared to that for the ionized sodium chloride. The trends in water flux and solute passage for both sodium chloride and boric acid as a function of hydraulic pressure were consistent with those generally predicted by reverse transport equations (i.e., Equations 1-4). Because of the analytical difficulties encountered with boric acid, the use of this solute was discontinued for the remaining experiments of this study.

Exposure of the membrane element to free chlorine at 1,000 (mg/L)×h resulted in an increase in water flux, but a decrease in solute passage. These changes do not appear to be consistent with membrane damage as a result of chemical attack because in such case the passage of both water and solute would be expected to increase. A possible explanation for the changes observed is that chlorine might have reacted with fouling material from prior use of the membrane element. A lower level of concentration polarization for the "cleaned" element would be consistent, at least in part, with the changes observed in membrane performance.

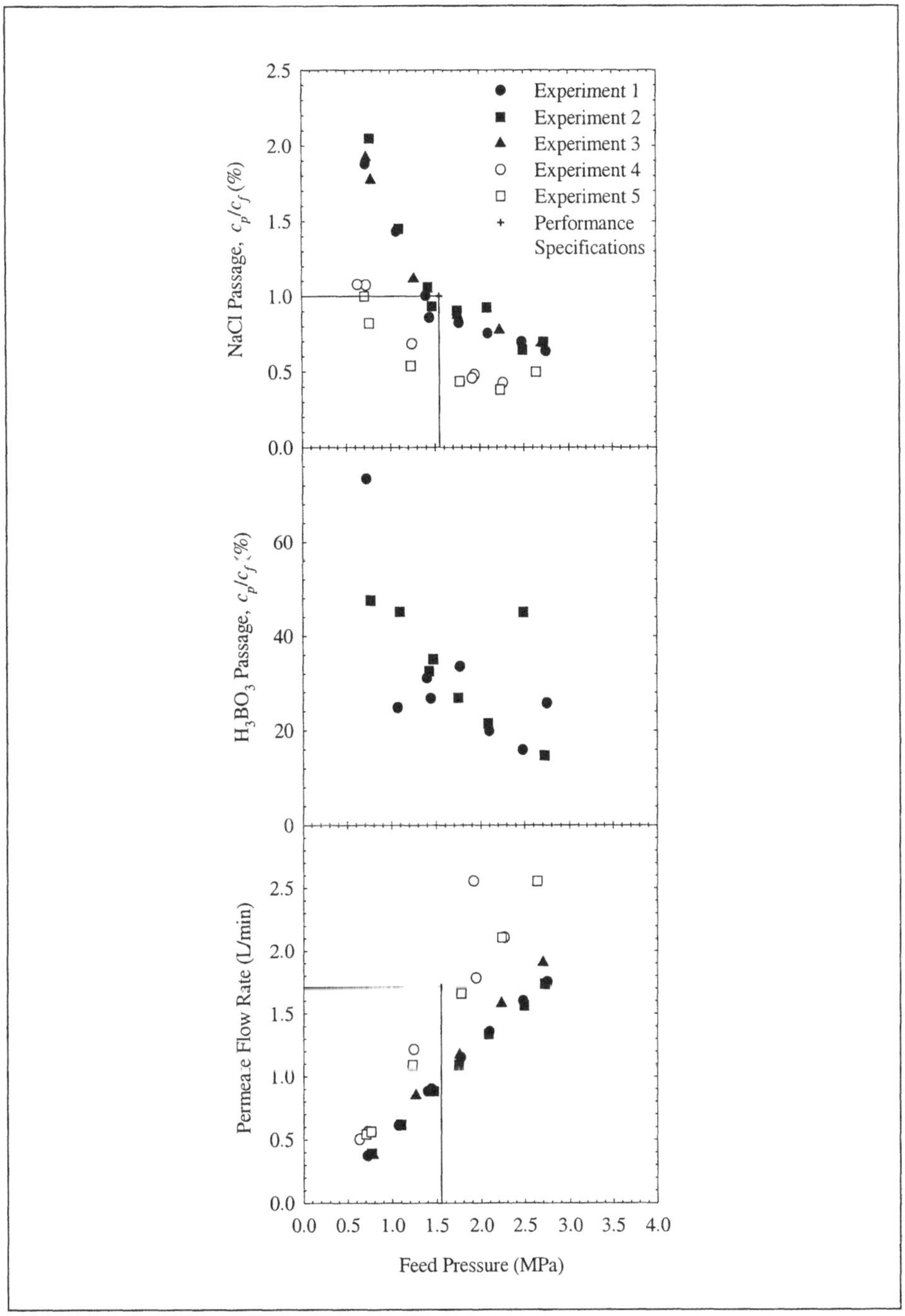

Figure 2.56 Performance of FilmTec BW30HP element before (Experiments 1-3), and after (Experiments 4-5) free chlorine exposure (see Table 2.5 for experimental conditions)

BW30HP - Virus Experiments

Feed Concentration Dynamics. Initial virus testing efforts with the FilmTec BW30HP element revealed that steady state conditions were not being reached with respect to the virus feed concentration. Experimental results for two experiments performed to assess the changes in virus concentration in the feed tank are presented in Figure 2.57. In the first seeding, the feed tank was seeded with a stock solution volume that resulted in an initial concentration of approximately 10^6 viruses/ml. The concentration of viruses dropped within the first 20 minutes of operation to approximately 10^2 viruses/ml, or a 4-log reduction. For the remaining 40 minutes of the testing period, the concentration leveled off. After testing for 1 hour, the feed tank was reseeded with stock solution at a higher initial concentration of approximately 10^7 viruses/ml. After the first 20 minutes of testing, the concentration of viruses only decreased by approximately 1 log. However, by the end of the second 60-minute period, the concentration of viruses was down to roughly 5.0×10^4 viruses/ml, or an overall reduction in excess of 3 logs. Two subsequent tests were conducted (data not shown in Figure 2.57), with similar results to the second seeding event. Therefore, a steady state was not being reached within the system. A possible explanation for these observations is that the viruses might be adhering to the various surfaces of the apparatus components. In order to ensure that the virus concentration in the feed tank remained at fairly constant levels during the time of testing, additional stock was injected into the tank prior to taking a sample.

Along with the feed samples that were obtained during the time test, permeate samples were also taken to verify whether viruses were passing through the membrane. No viruses were found in any of the permeate samples, as shown in Figure 2.58. Therefore, viruses were being rejected at levels as high as 7-log with the FilmTec BW30HP element prior to chlorine exposure.

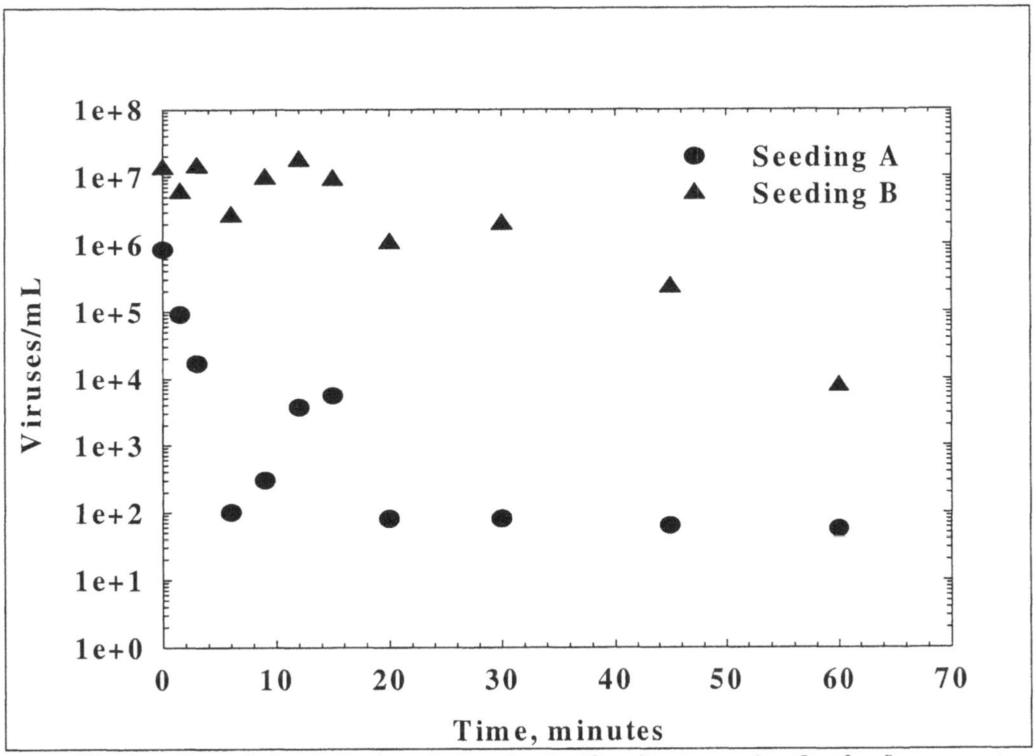

Figure 2.57 Results of time testing to evaluate the decrease in the feed concentration of viruses

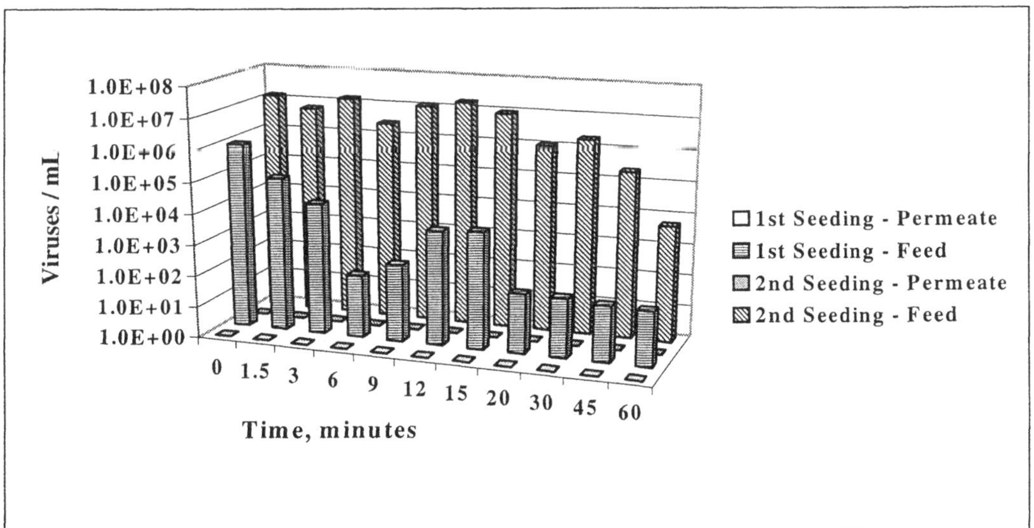

Figure 2.58 Virus removal observed during time test (Experiment 3) with the FilmTec FT30 element

Assessment of virus removal with FilmTec BW30HP element exposed to free chlorine. Experiment 4 was designed to assess the rejection of viruses by the FilmTec BW30HP element after exposure to 1,000 (mg/L)xh of free chlorine. Virus experimental results for Experiment 4 are presented in Figure 2.59. Five data points were obtained as a function of the hydraulic pressures indicated in the figure, which corresponded to permeate fluxes in the range of 5.4 to 27.4 gfd. The feed virus concentration ranged from 1.0×10^6 to 8.0×10^7 viruses/ml. As depicted in the figure, no virus passage was observed confirming that the exposure to free chlorine did not compromise the integrity of the membrane element. The virus removal for this membrane was greater than 6 to 8 logs even though chlorine exposure caused membrane damage as evidenced by the impact on water and solute transport discussed previously.

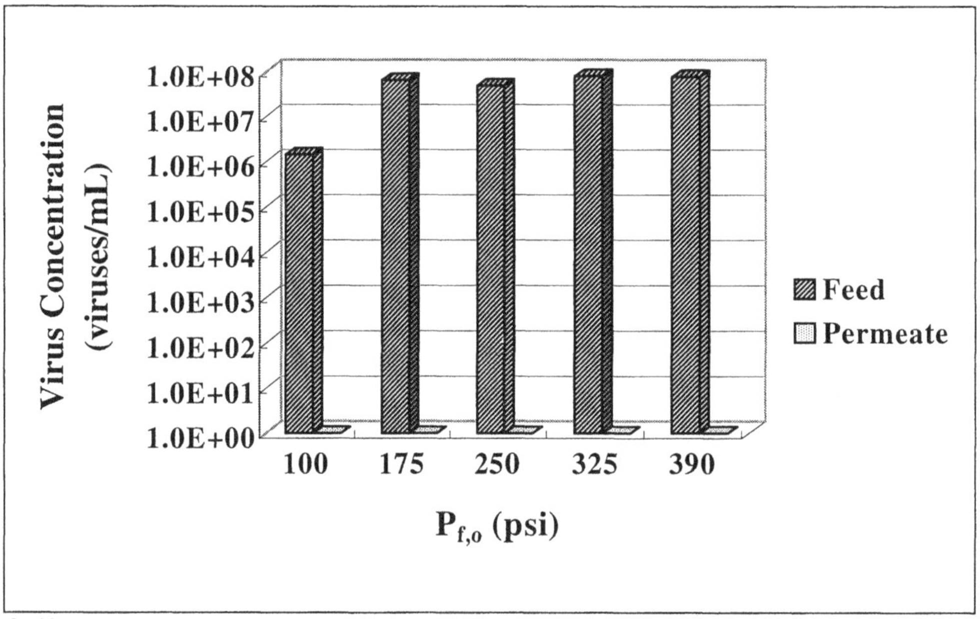

Figure 2.59 Virus removal observed for Experiment 4 performed with the FilmTec FT30 element

ESPA1 - Solute Experiments

Water fluxes and sodium chloride passages observed during Experiments 1 and 2 with the Hydranautics ESPA1 element are presented in Figure 2.60. In general, the experimental results obtained were generally consistent with performance specifications provided by the membrane manufacturer, which are also indicated in Figure 2.60. Notice that the specified solute passage corresponds not to a nominal but to a minimum solute rejection value (Table 2.4). The slightly greater value for the water flux specification is also consistent with the membrane performance being provided for a more dilute solution (1,500 mg/L versus 1,800 mg/L).

ESPA1 - Virus Experiments

Feed Concentration Dynamics. The stability of virus concentration in the feed tank after a seeding event was assessed with the Hydranautics ESPA1 element as part of Experiment 3. Experimental results obtained are presented in Figure 2.61. As shown in the graph the feed concentration of viruses decreased from an initial value of 1.2×10^8 to 1.4×10^2 within 45 minutes. These results are similar to those dicussed previously for the FilmTec BW30HP membrane. However, in contrast to the BW30HP results, virus passage was observed during the first 6 min of testing with the ESPA1 element. The corresponding rejection was approximately 7 logs and therefore once the feed concentration of viruses dropped below 10^7 viruses/ml, the permeate concentrations were found to be below detection.

Effect of Water Flux. Additional tests were performed as part of Experiment 3 to assess the effect of water flux on virus passage. Virus stock solution was added approximately four minutes before sampling at each pressure tested. Experimental results for these tests are presented in Figure 2.62 in terms of feed and permeate concentrations of viruses versus feed pressure. The virus concentration achieved in the feed tank ranged from 6.1×10^6 to 1.2×10^8, and the permeate concentrations corresponded to virus rejections of 5.3 to 6.4 logs. These rejections appeared to be independent of water flux consistent with the occurrence of advective passage through a small pinhole.

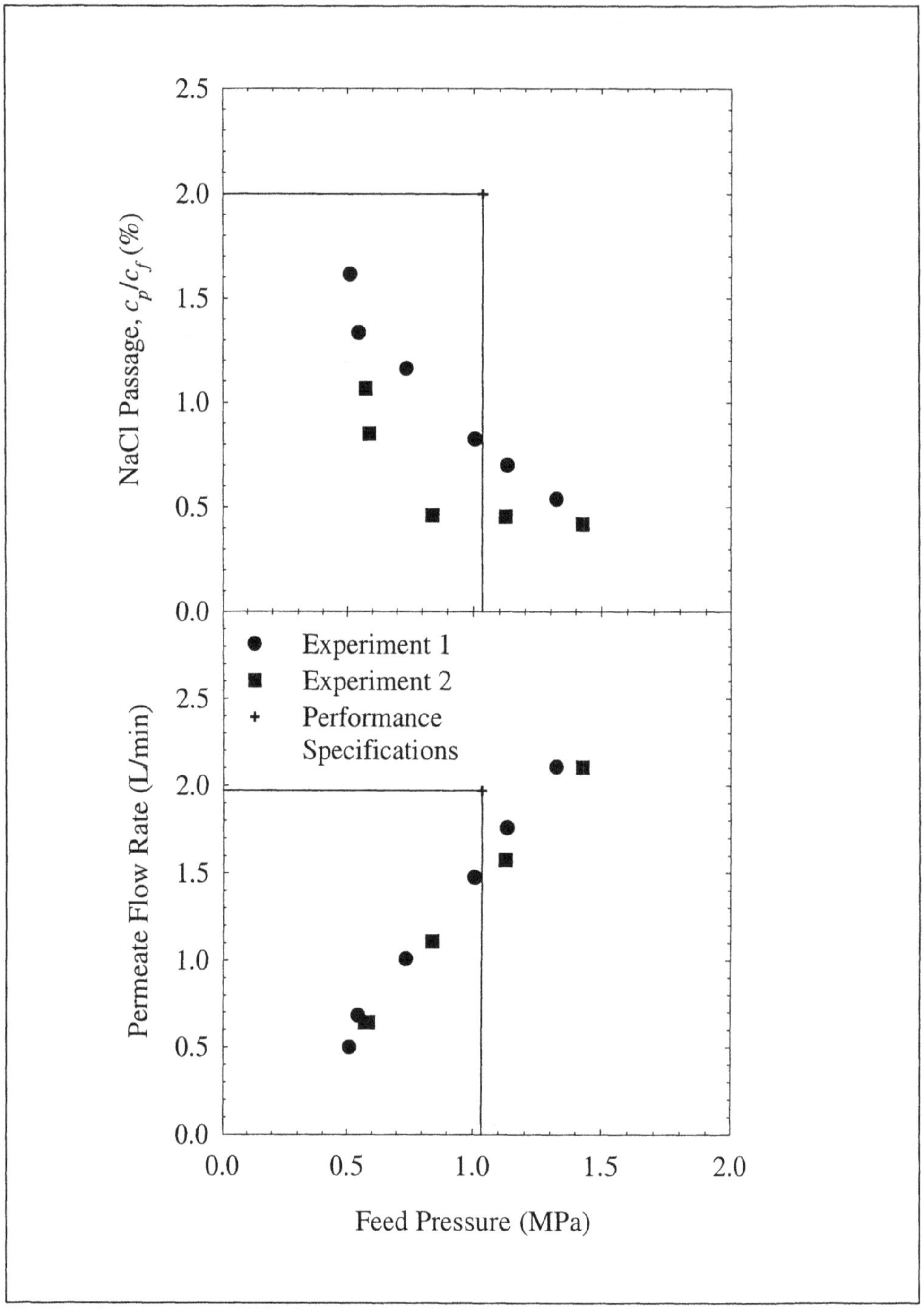

Figure 2.60 Performance of Hydranautics ESPA1 element before virus addition (see Table 2.6 for experimental conditions)

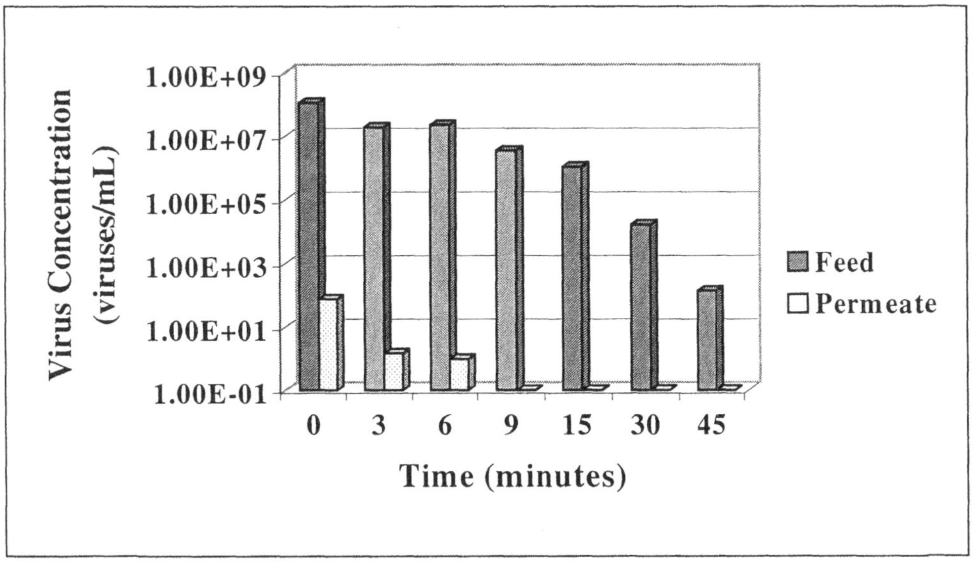

Figure 2.61 Virus removal observed during time test (Experiment 3) with the Hydranautics ESPA1 element

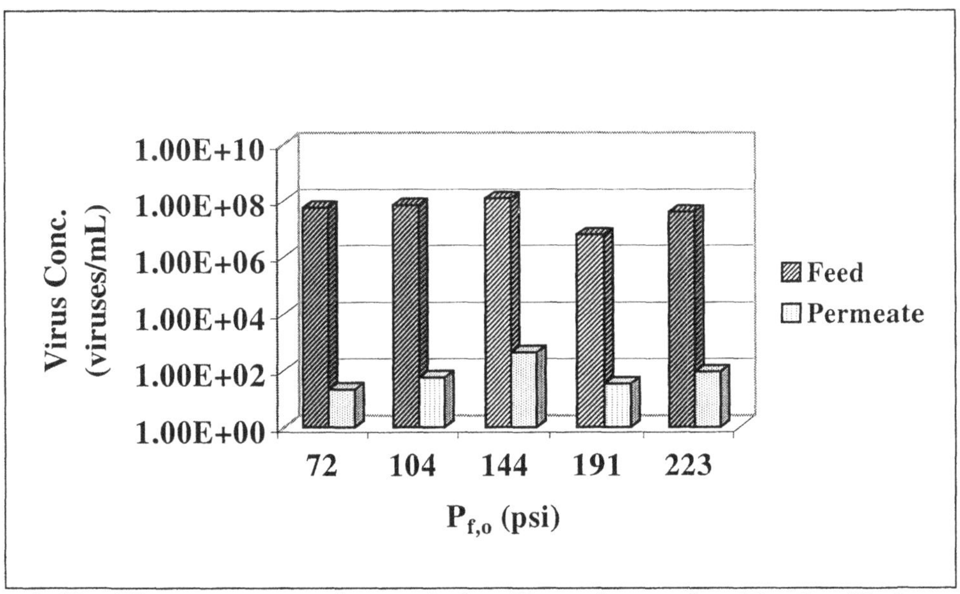

Figure 2.62 Virus passage observed for Experiment 3 with the Hydranautics ESPA1 element

ESPA1 - Virus/Microsphere Experiments

The objective of Experiment 4 was to correlate the passage of viruses and microspheres with the Hydranautics ESPA1 element. The initial addition of microsphere stock solution resulted in a feed concentration of approximately 10 mg/L. However, it was assumed that, similar to the phenomena observed for viruses, the concentration of microspheres in the feed tank would decrease over time. Hence, additional microsphere dosages of 10 mg/L were added before each sampling event. A stabilization period of only 1 hour was allowed between samples. The results for Experiment 4 are shown in Figures 2.62 and 2.63 for viruses and microspheres, respectively.

Feed virus concentrations ranged from 8.0×10^{8} to 1.8×10^{9} and the corresponding rejections ranged from 6.7 to 7.9 logs without a trend with water flux in accordance with Experiment 3 results discussed previously.

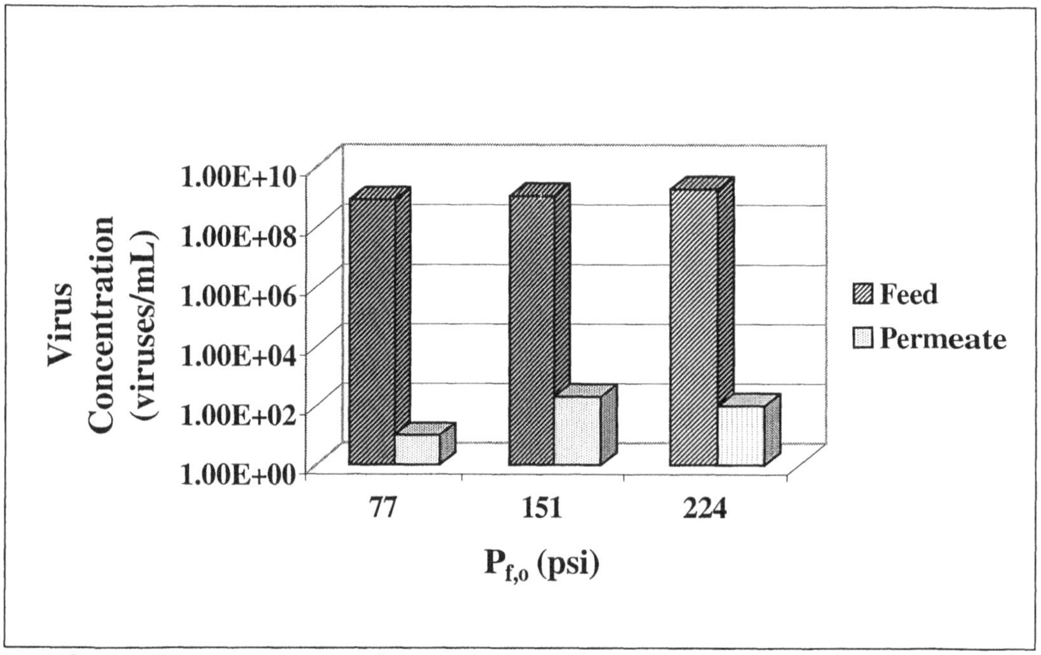

Figure 2.63 Virus passage observed for Experiment 4 performed with the Hydranautics ESPA1 element

Experiment 4 results for microspheres are presented in Figure 2.64. The fluorescence of all permeate samples was found to be below the detection limit somewhere in the range of 0.01-0.1 μg/L. Consequently, the removal of microspheres was in excess of 5-6 logs. Although these results are consistent with the observations for viruses, a direct correlation could not be established due to analytical detection limitations for the microspheres.

ESPA1 - MS2 Phage/Microsphere Experiments with O-ring Imperfection

Experiment 5 was performed to assess if a faulty o-ring from the seal between the permeate tube and the vessel end-cap would affect the performance of the system. A small crack induced in an o-ring for this purpose is depicted in Figures 2.65 and 2.66 (before and after crack induction, respectively). The Hydronautics ESPA1 membrane element was tested installed with this o-ring for virus and microsphere rejection.

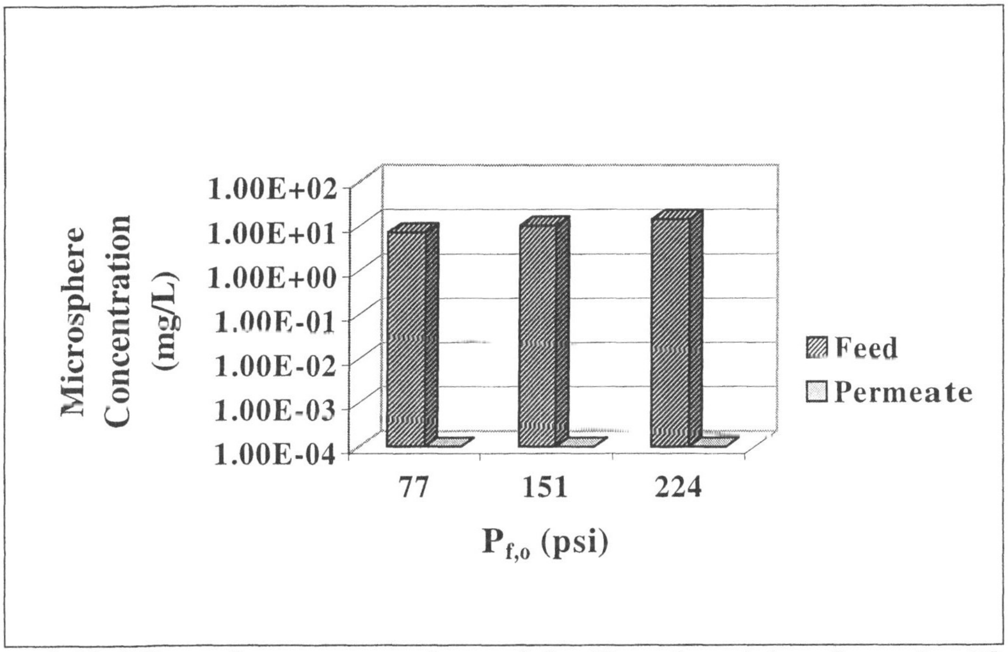

Figure 2.64 Microsphere passage observed for Experiment 4 performed with the Hydranautics ESPA1 element

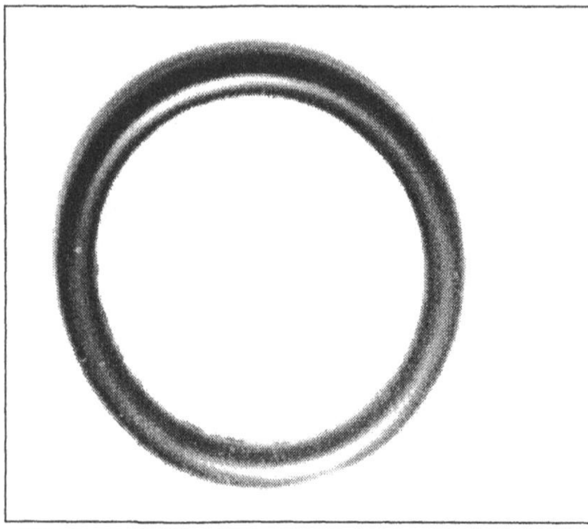

Figure 2.65 Intact o-ring tested with the Hydranautics ESPA1 element

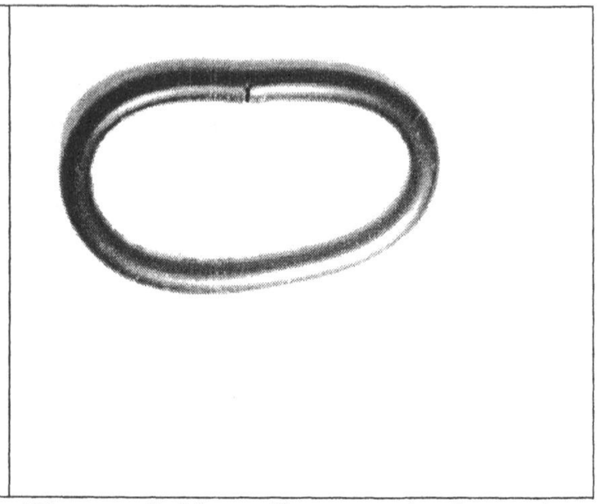

Figure 2.66 Compromised o-ring tested with the Hydranautics ESPA1 element

Experimental results for this test are presented in Figures 2.67 and 2.68 for viruses and microspheres, respectively. The virus concentration in the feed was $1.9\text{-}2.6 \times 10^8$, and those in the permeate corresponded to 6.2-7.0 logs removal, generally consistent with previous results obtained with this element.

Microsphere removals in the order of 5-6 logs are presented in Figure 2.68, but the permeate concentration are within the detection limit range of 0.1 to 0.01 μg/L. Consequently, the microsphere results should be interpreted as corresponding to rejection levels in excess of 5 logs.

Both viruses and microspheres were rejected at approximately the same levels as those found with the same membrane element prior to inducing the crack in the o-ring. Apparently the cracked o-ring still served as an effective seal and barrier agents the passage of particles greater than approximately 20 nm.

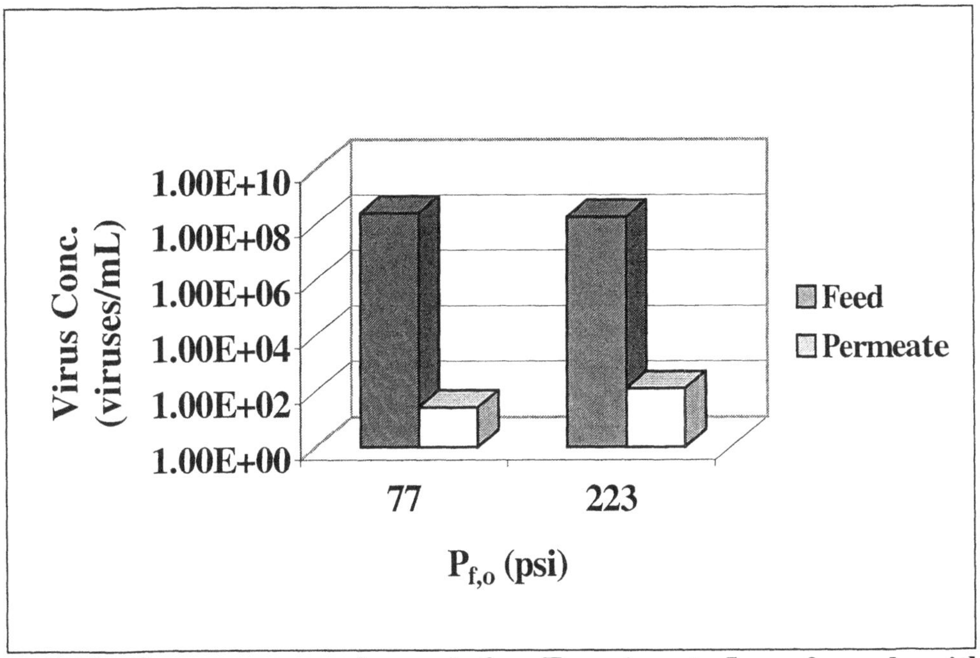

Figure 2.67 Virus passage observed for Experiment 5 performed with the Hydranautics ESPA1 element installed with a compromised o-ring

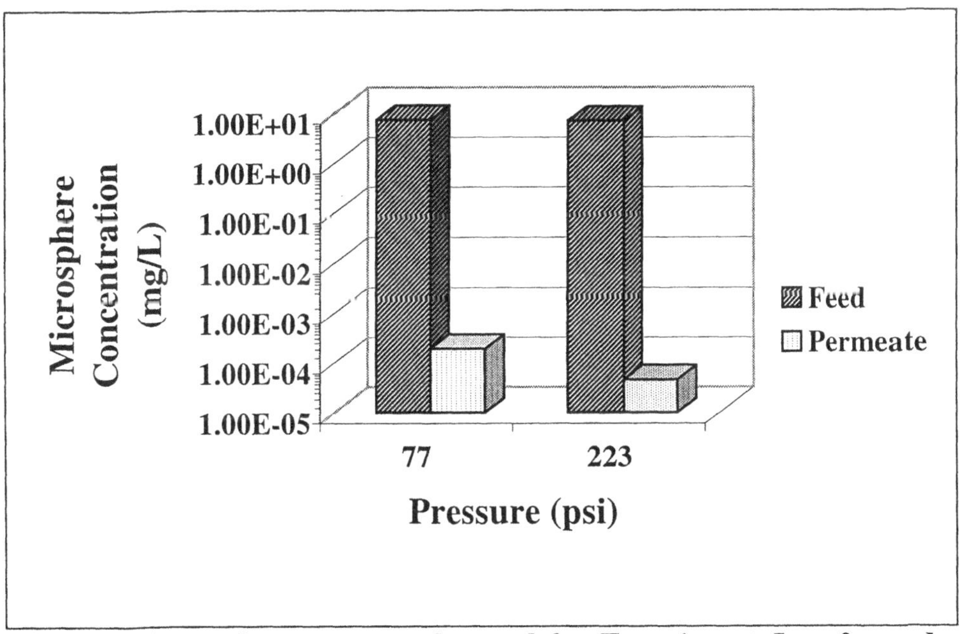

Figure 2.68 Microsphere passage observed for Experiment 5 performed with the Hydranautics ESPA1 element installed with a compromised o-ring

ROGA - Solute Experiments

Water and solute passage results for all experiments performed with the Fluid Systems ROGA-HR membrane element are presented in Figure 2.69. Membrane performance before inducing the imperfections was the same for the three experiments performed and consistent with performance specifications. The small pinhole resulted in slight increase in solute passage without noticeable changes in water permeability (Experiments 4 and 5 in Figure 2.69). The larger pinhole resulted in more than double passage of sodium chloride consistent with a corresponding increase in water permeability of approximately 6-7 percent.

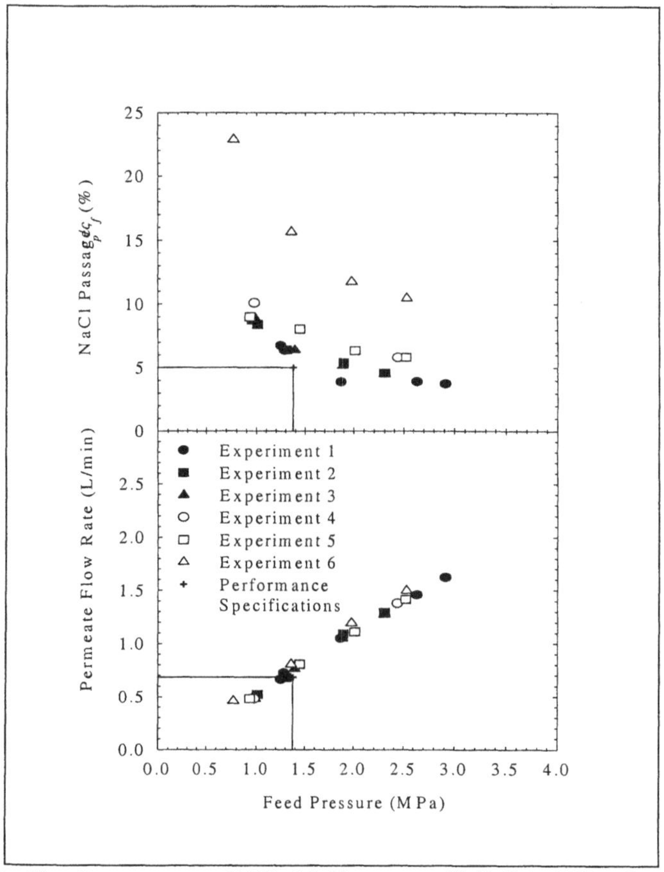

Figure 2.69 Performance of Fluid Systems ROGA-HR element before (Experiments 1-3), and after inducing relatively small (Experiments 4-5) and large (Experiment 6) pinholes (see Table 2.7 for experimental conditions)

Feed Concentration Dynamics. Experiments 1 and 2 included tests designed to assess the rate of virus concentration decline in the feedwater over time. Two separate virus seedings were done; however, they were not consecutive. Both were performed after the tank had been filled with a fresh batch of feedwater. The results from the time tests are shown in Figure 2.70. Both seedings A and B demonstrated consistent decline in the concentration from the initial feed concentration of approximately 4×10^5-3×10^6 viruses/ml. After roughly 20 minutes, the concentration had decreased nearly 5 logs. This information was used in determining the sampling protocol for the virus testing at variable pressure discussed in the next section. Virus concentrations for all time test permeate samples were below the detection limit of 1 virus/ml.

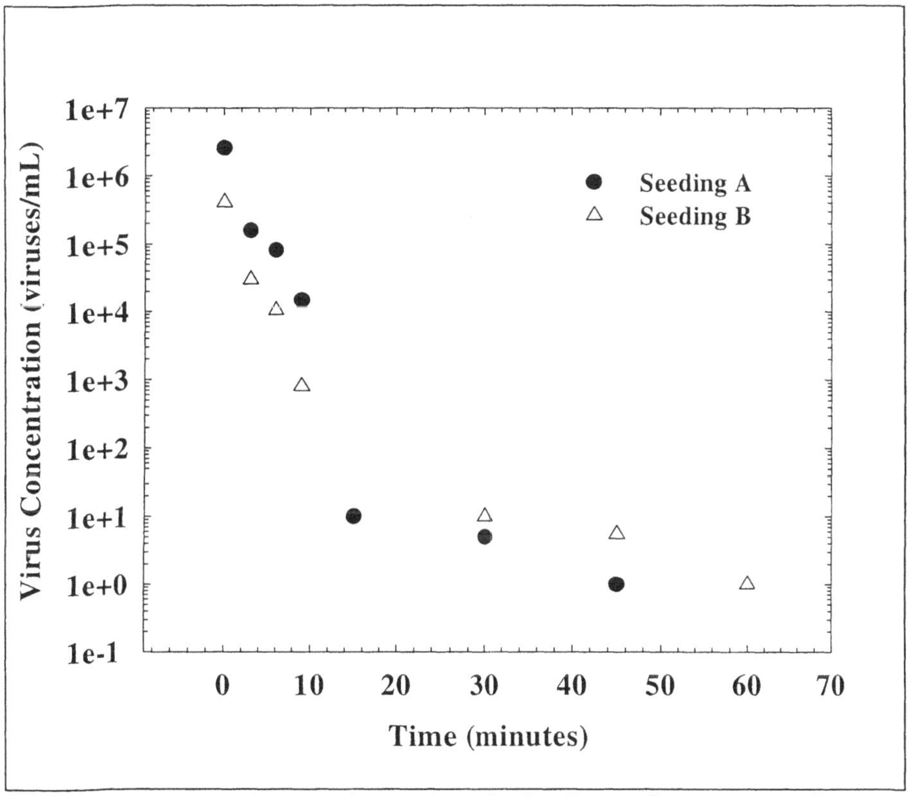

Figure 2.70 Virus time testing performed with the Fluid Systems ROGA-HR element as part of Experiments 1 (Seeding A) and 2 (Seeding B)

ROGA-HR - MS2 Phage Experiments

Virus passage results obtained with the Fluid Systems ROGA-HR element before inducing the pinholes are presented in Figures 2.71, 2.72, and 2.73 for Experiments 1, 2, and 3, respectively. No microspheres were added for these tests. Some difficulties were encountered in maintaining a constant feed concentration for Experiment 1 as depicted in Figure 2.71. Nevertheless removals in excess of 4.8 logs were achieved at the higher pressures investigated.

Feed virus concentrations were more constant at 1.0-1.6×10^5 viruses/ml for Experiment 2 as depicted in Figure 2.72. Once again, virus rejections greater than approximately 5 logs were observed.

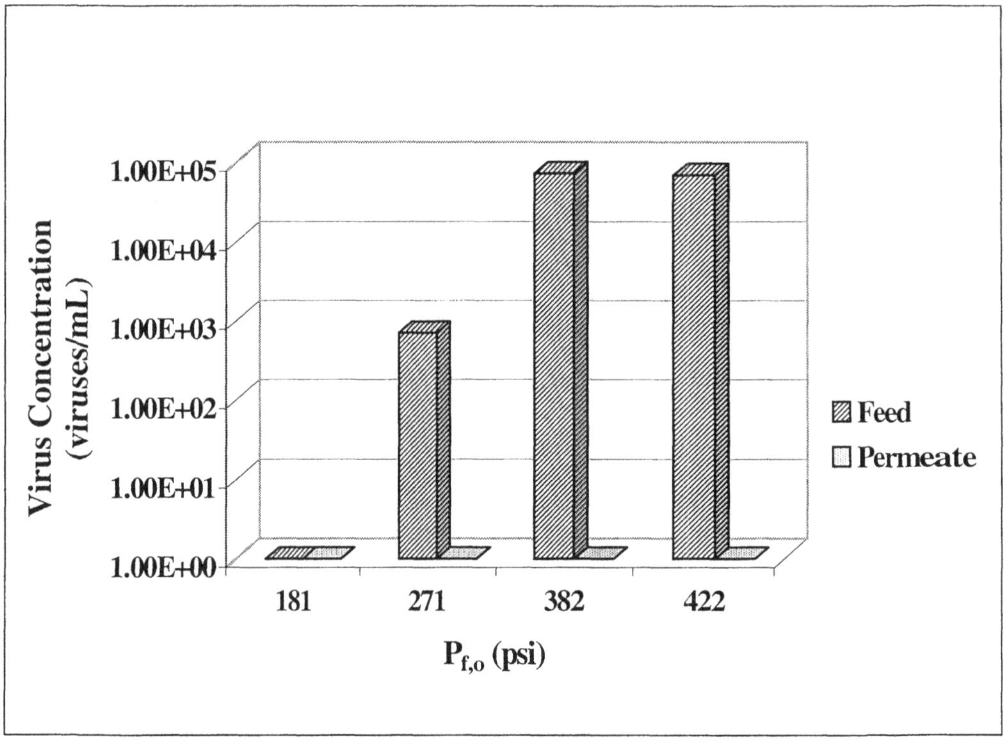

Figure 2.71 Virus removal observed for Experiment 1 with the Fluid Systems ROGA-HR element

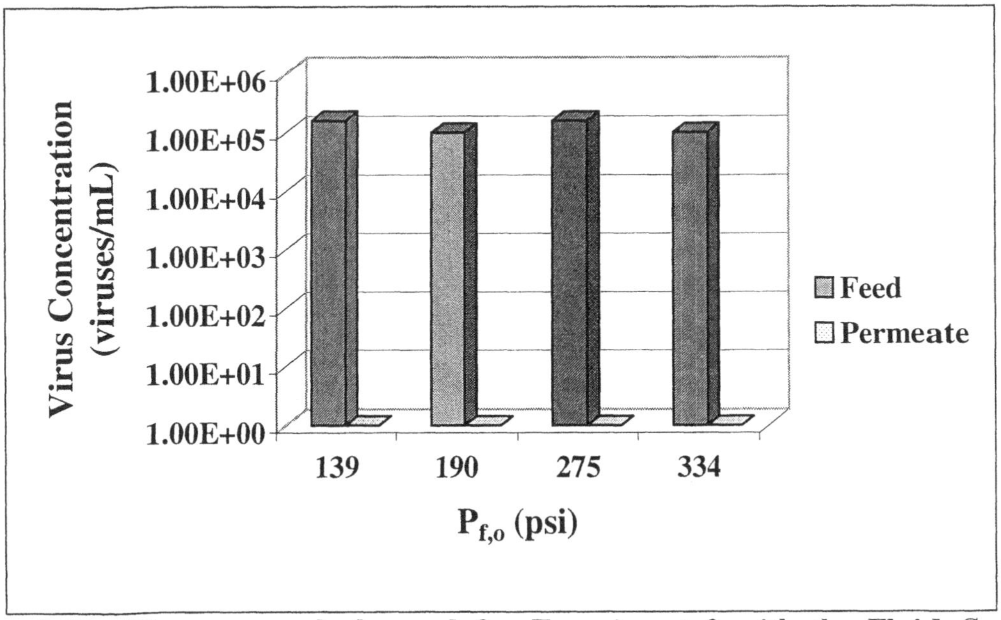

Figure 2.72 Virus removal observed for Experiment 2 with the Fluid Systems ROGA-HR element

Experiment 3 was performed to assess if higher log removals were achieved with the Fluid Systems ROGA-HR element. Experimental results are presented in Figure 2.73. The feed virus concentrations ranged from 3.0×10^6 to 7.5×10^6 viruses/ml, and once again no viruses were detected in permeate. Therefore, removals achieved with this element prior to inducing imperfections were in excess of 6.5 logs.

Characterization of Induced Pinholes. The small and large imperfection induced in the surface of the Fluid Systems ROGA-HR element for Experiments 4-6 described subsequently, were characterized with an Olympus BX60 (Leco Corportation, Chicago, IL) microscope and a CH250 Photometric Charge Captured Device with a Kodak KAF 1400 Chip (Photometrics, AZ) at a magnification of 40 times. The size of the small and large pinholes were estimated at approximately 10 and 100 μm, respectively.

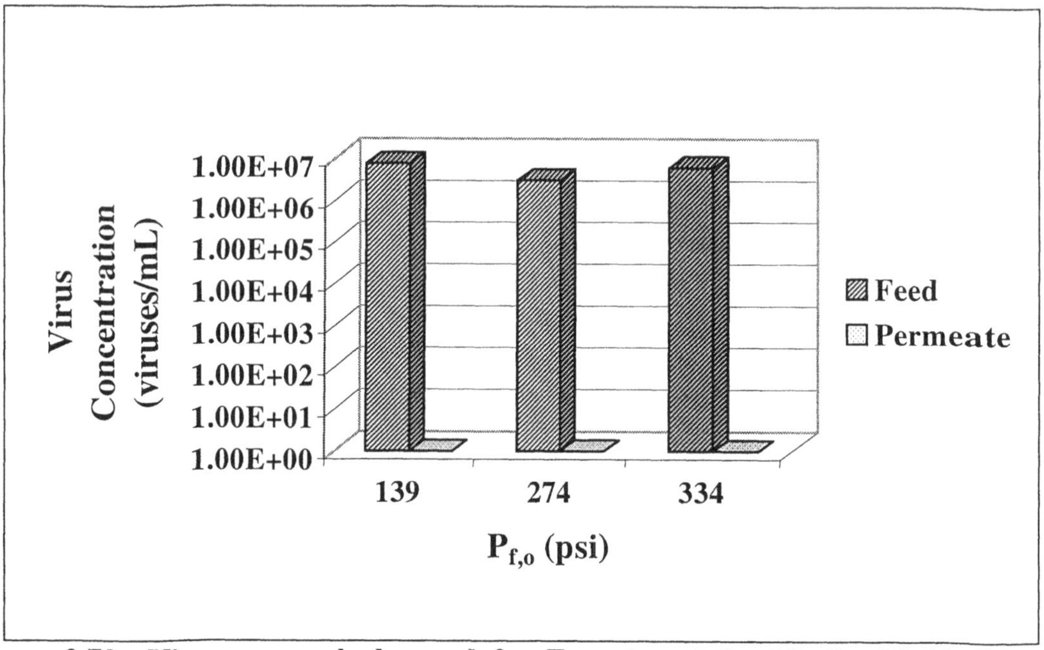

Figure 2.73 Virus removal observed for Experiment 3 with the Fluid Systems ROGA-HR element

ROGA-HR - MS2 Phage/Microsphere Experiments with Induced Pinholes

The Fluid Systems ROGA-HR element was first tested after inducing relatively small and large pinholes in order to attempt the development of a correlation between viruses and microspheres at measurable microsphere rejection levels. Experiment 4 results for viruses and microspheres are presented in Figures 2.74 and 2.75, respectively. As depicted in the figure, there was good agreement between removals for viruses, 2.4-2.8 logs, and microspheres, 3.0 logs. Furthermore, no dependence of particle passage on water flux was observed indicating that the passage is consistent with advection through the induced pinholes.

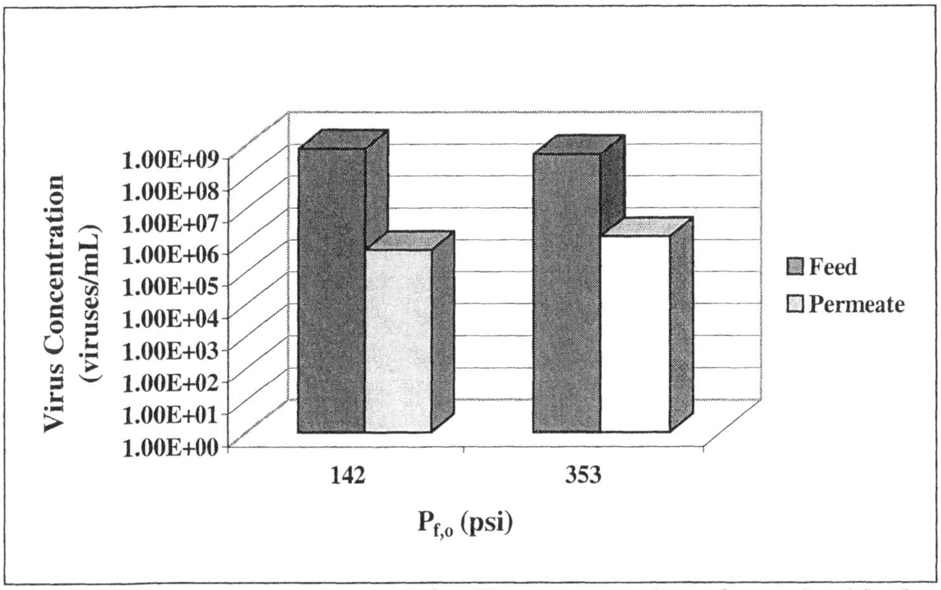

Figure 2.74 Virus passage observed for Experiment 4 performed with the Fluid Systems ROGA-HR element compromised with a small pin-sized hole

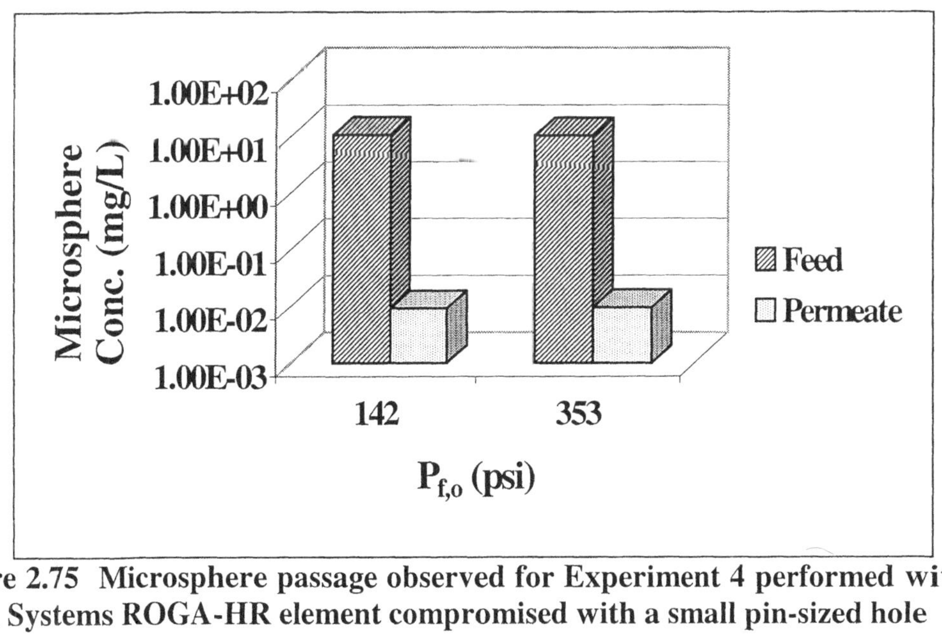

Figure 2.75 Microsphere passage observed for Experiment 4 performed with the Fluid Systems ROGA-HR element compromised with a small pin-sized hole

Experiment 5 was performed to assess if passage of viruses and microspheres through the small pinhole would change within a period of 24 hours. The results for Experiment 5 are presented in Figures 2.76 and 2.77 for viruses and microspheres, respectively. As can be seen from the figures, the removal of both viruses, 6.6-7.6 logs, and microspheres, >4.7-5.1 logs, increased considerably compared to those in Experiment 4. A possible explanation for the increase in removal was that the particles and perhaps organic matter from the virus stock were clogging the small imperfection causing a healing effect to occur in the membrane.

Experiment 6 was performed with the same Fluid Systems ROGA-HR element after inducing a second pinhole approximately ten times larger than the previous imperfection. Again, the element was tested with viruses and microspheres simultaneously. The experimental results are presented in Figures 2.78 and 2.79 for viruses and microspheres, respectively. Once again, good agreement was found between removals for viruses, 1.0-1.6 logs, and microspheres, 0.9-1.2 logs.

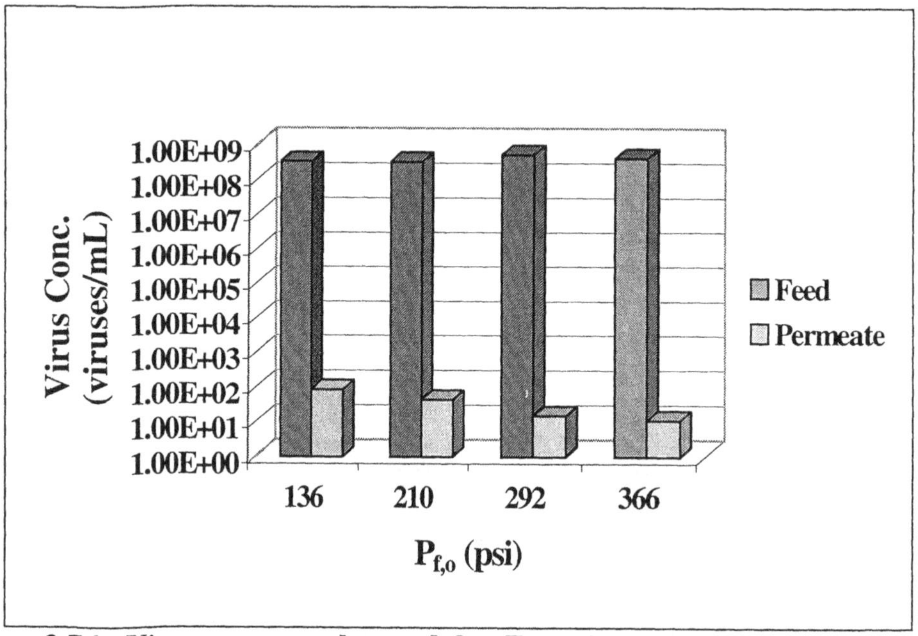

Figure 2.76 Virus passage observed for Experiment 5 performed with the Fluid Systems ROGA-HR element compromised with a small pin-size hole

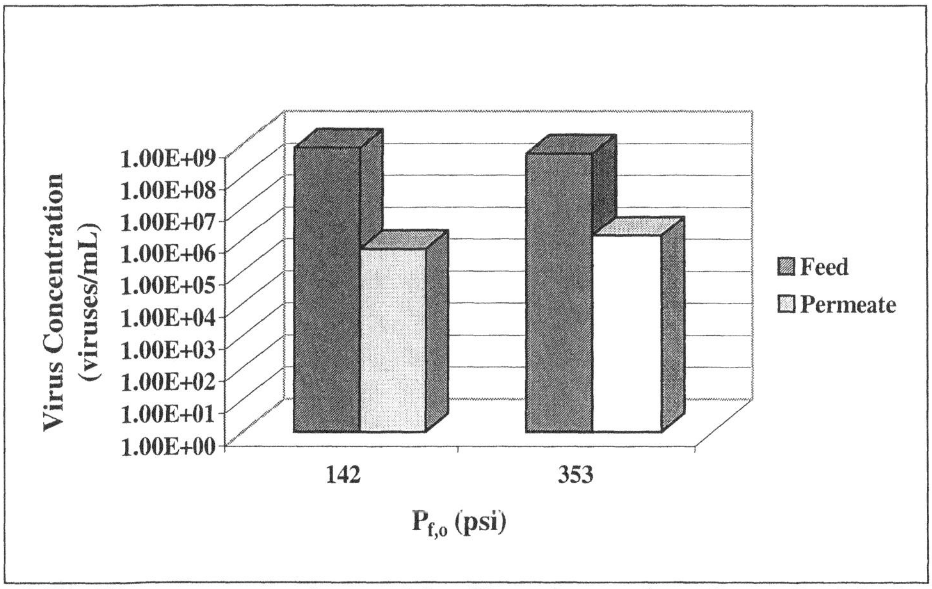

Figure 2.74 Virus passage observed for Experiment 4 performed with the Fluid Systems ROGA-HR element compromised with a small pin-sized hole

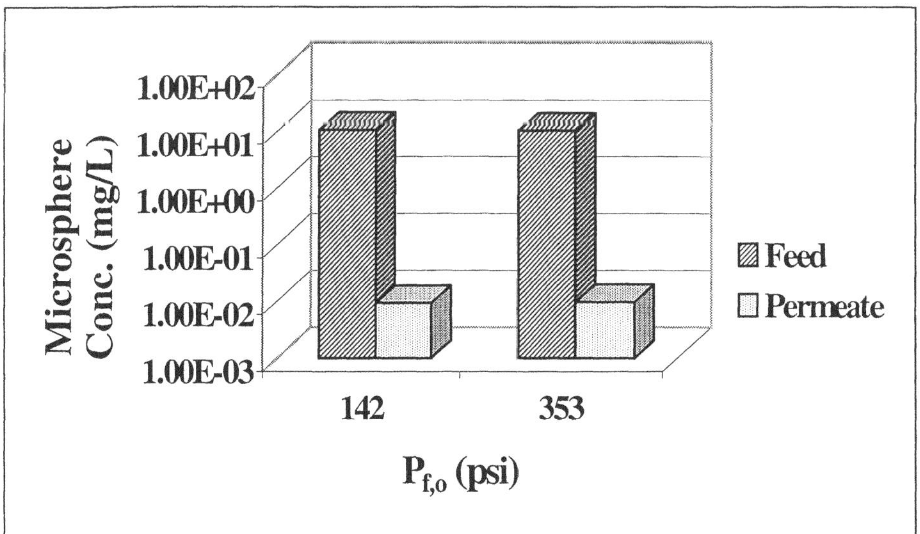

Figure 2.75 Microsphere passage observed for Experiment 4 performed with the Fluid Systems ROGA-HR element compromised with a small pin-sized hole

Experiment 5 was performed to assess if passage of viruses and microspheres through the small pinhole would change within a period of 24 hours. The results for Experiment 5 are presented in Figures 2.76 and 2.77 for viruses and microspheres, respectively. As can be seen from the figures, the removal of both viruses, 6.6-7.6 logs, and microspheres, >4.7-5.1 logs, increased considerably compared to those in Experiment 4. A possible explanation for the increase in removal was that the particles and perhaps organic matter from the virus stock were clogging the small imperfection causing a healing effect to occur in the membrane.

Experiment 6 was performed with the same Fluid Systems ROGA-HR element after inducing a second pinhole approximately ten times larger than the previous imperfection. Again, the element was tested with viruses and microspheres simultaneously. The experimental results are presented in Figures 2.78 and 2.79 for viruses and microspheres, respectively. Once again, good agreement was found between removals for viruses, 1.0-1.6 logs, and microspheres, 0.9-1.2 logs.

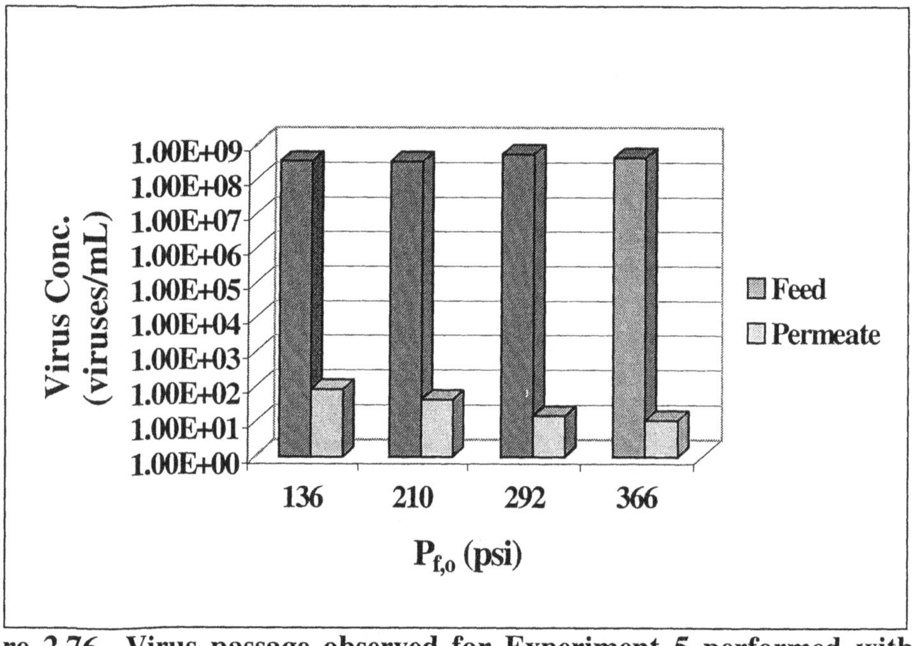

Figure 2.76 Virus passage observed for Experiment 5 performed with the Fluid Systems ROGA-HR element compromised with a small pin-size hole

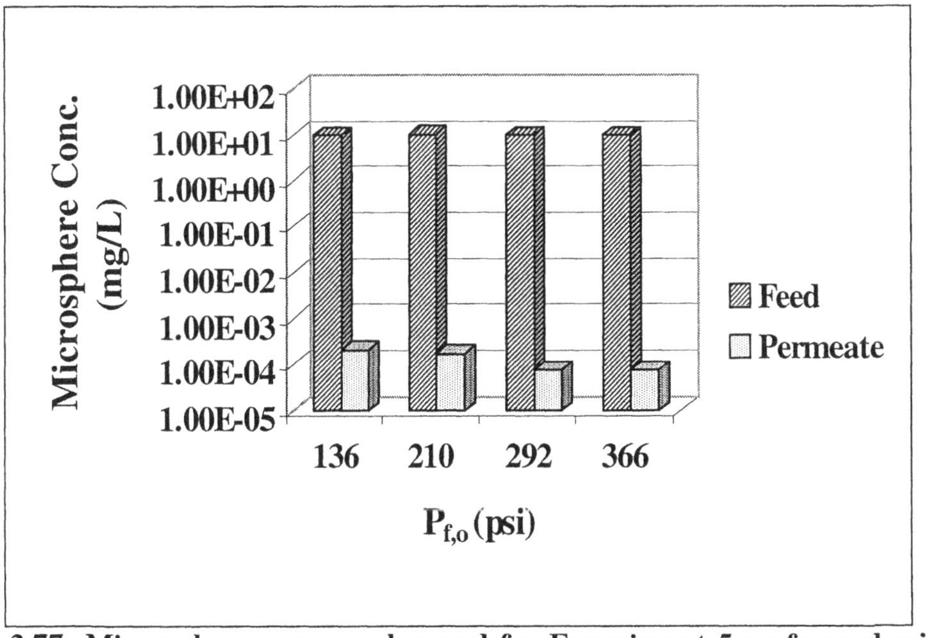

Figure 2.77 Microsphere passage observed for Experiment 5 performed with the Fluid Systems ROGA-HR element compromised with a small pin-size hole

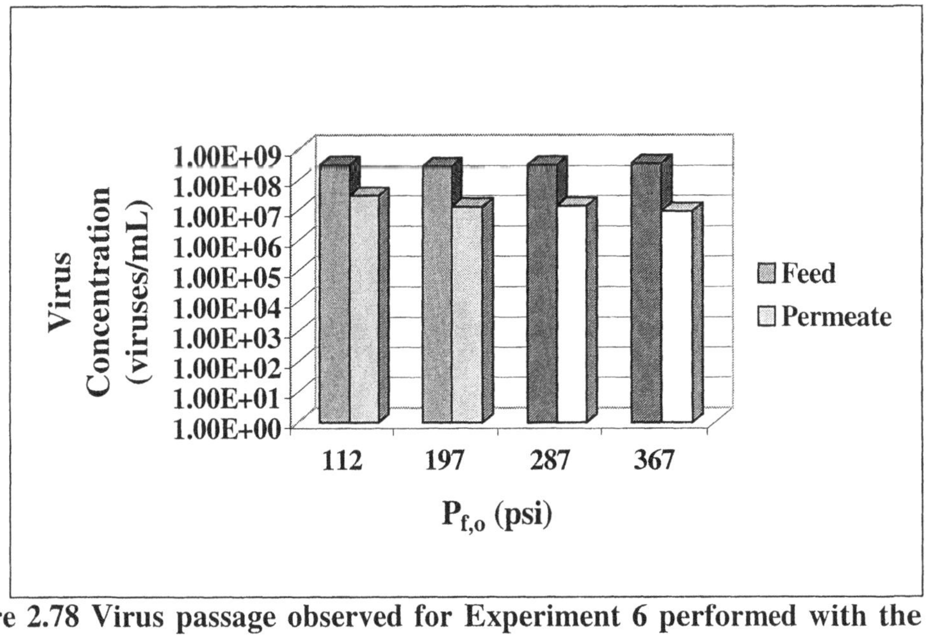

Figure 2.78 Virus passage observed for Experiment 6 performed with the Fluid Systems ROGA-HR element compromised with a large pinhole

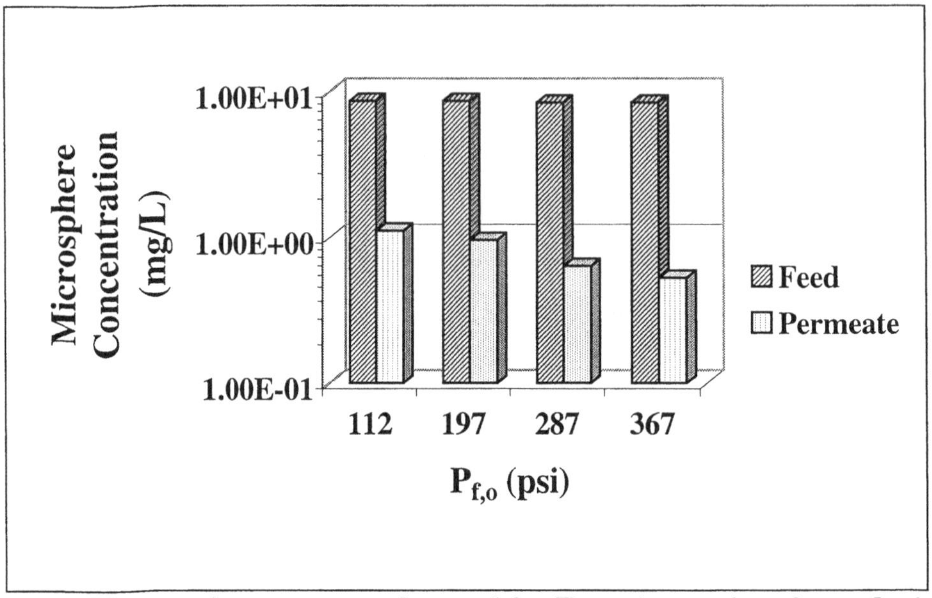

Figure 2.79 Microsphere passage observed for Experiment 6 performed with the Fluid Systems ROGA-HR element compromised with a large pinhole

CHAPTER 3
PILOT-SCALE TESTING

INTRODUCTION

This section of the report presents the pilot-scale testing conducted at the following sites:

- Orange County Water District (OCWD) Water Factory 21 (WF-21) in Fountain Valley, CA
- The Northwest River Water Treatment Plant (NWR-WTP) in the City of Chesapeake, VA

The feed waters to the test elements in these two sites were as following:

- Synthetic water, i.e., permeate from full-scale surface water RO facility at NWR-WTP
- Microfiltered secondary effluent (MFSE) at WF-21
- Conventionally treated river water (CTRW) polished using 10-μm cartridge filters at the NWR-WTP

Waters from WF-21 and NWR-WTP were selected to provide appropriate matrices for evaluating the project objectives. Due to the more stringent forthcoming regulations, microbial removal by high-pressure membranes is becoming essential when treating surface waters for drinking water use and wastewater effluent for water reuse applications. High levels of microbial removal from surface waters (and groundwaters under the influence of surface waters) is dictated by the requirements of the Enhanced Surface Water Treatment Rule (ESWTR). Although there are no federal regulations regarding disinfection of wastewater effluent, many states, most notably California, Georgia and Florida, have promulgated or proposed regulations requiring pathogen control, particularly where the reclaimed effluent is to be used for indirect potable reuse.

OBJECTIVES

The main objective of this portion of the project was to evaluate, on a pilot-scale level, the removal of enteric virus surrogates by intact or compromised RO and NF membranes and intact or compromised o-rings using both microbial and non-microbial surrogate indicators. The selected microbial surrogate was MS2 phage due to its similar size, shape, and nucleic acid composition to enteric viruses. The non-microbial surrogate indicators were 0.02-μm orange fluorescent carboxylate-modified microspheres and Rhodamine WT (R-WT), a fluorescent dye. R-WT, not evaluated in the bench-scale portion of the project, was added prior to the start of pilot testing based on advancements in fluorometer design that occurred between the start of the project (January 1998) and the planning of the pilot testing (June 2000), and now allow for detection of R-WT at levels as low as 10 to 20 ng/L in clean water. This would allow a challenge level of 4-5 logs at feed concentration of 0.1 to 1 mg/L, levels that are approved for use by the EPA with water treatment facilities. R-WT has a molecular weight of approximately 480, and, as such, should be very highly rejected by high-pressure membranes.

The specific objectives were to:

- Confirm the removal of virus, microsphere and R-WT by intact RO and NF membranes previously tested in bench-scale experiments using both flat-sheet coupons and single, small capacity spiral-wound elements;
- Determine the impact of RO and NF membrane compromization (through induced imperfections penetrating both the active and support layers) on virus, microsphere and R-WT removals;
- Determine the impact of RO and NF membrane fouling and further chemical cleaning on virus, microsphere and R-WT passage through compromised RO and NF membranes;
- Determine the level of o-ring compromization to result in the increased passage of virus, microsphere and R-WT;
- Assess, by visual means, the reactivity of RO and NF elements towards the uptake of R-WT;
- Determine the impact on virus, microsphere and R-WT removal of a compromised RO element or o-ring when installed at different locations within a multiple-array (stage) membrane system;
- Determine the relationship between the rejections of the microbial- and non-microbial virus surrogate indicators with that of the standard indicator of RO and NF membrane integrity, conductivity.

MATERIALS AND METHODS

Description of the Test Sites

Water Factory-21

WF-21 is a 15-mgd advanced water reclamation facility that treats secondary effluent from the adjacent Orange County Sanitary District wastewater treatment plant using high-pH lime clarification, recarbonation, granular media filtration, RO and granular activated carbon for direct aquifer injection to control coastal seawater intrusion. Further, WF-21 contains three demonstration-scale advanced treatment trains, consisting of continuous-flow MF followed by RO, that also treat secondary effluent. These facilities have been put in place to demonstrate the efficacy and cost effectiveness of MF as an alternative to lime-based treatment used at the 15-mgd facility. OCWD is currently planning the construction of a 75-mgd MF/RO facility (including ultraviolet disinfection) as a means to increase the production of high quality reuse water for groundwater replenishment. This large scale facility reflects the growing importance in the use of MF and RO for advanced reclamation, not only in California, but in other water supply challenged regions in both the U.S and the world. High levels of microbial removal are dictated by the requirements of indirect potable reuse regulations that have been proposed or promulgated by individual State regulatory agencies (e.g., California Department of Health Services). As such, the use of MF-treated secondary effluent was chosen as one of the water matrices for the pilot scale membrane integrity testing.

Northwest River Water Treatment Plant

The 10-mgd NWR-WTP treats water from the Northwest River in Virginia. The Northwest River receives input from the Great Dismal Swamp, a water body containing very high levels of natural organic matter (typical total organic carbon levels of 50 mg/L) and is tidally influenced by seawater intrusion from the Carrituck Sound. Typically, the water quality of the Northwest River is characterized by low TDS, alkalinity and hardness, moderate turbidity and high TOC and disinfection by-product precursors, however during low-flow periods, chloride and TDS levels can exceed 1,500 and 3,000 mg/L, respectively. The NWR-WTP uses conventional treatment processes: coagulation with alum and lime, flocculation, sedimentation and dual-media granular filtration. A portion of the filter effluent is further treated by four RO trains for salinity and TOC reduction (for control of chlorinated byproducts in the finished water. When salinity (and chlorides) are low, the portion of filter effluent treated by RO treatment is minimized to that required to maintain finished water DBP levels below regulatory limits. Additionally, the plant also contains separate RO trains that treat highly brackish groundwater.

Test Units

WF-21.

Two units were employed for testing in WF-21: a dual-element test unit (DETU) (Figure 3.1) and a multi-stage test unit (MSTU) (Figure 3.3). Figures 3.2 and 3.4 show photos of the DETU and MSTU units, respectively.

The DETU consisted of two pressure vessels arranged in series along with associated instrumentation and controls. Each vessel was capable of holding a single 4-inch diameter by 40-inch long membrane element. For these experiments, the ESPA-1 RO element was operated in lead vessel (referred to hereafter as the RO vessel; vessel No.1 in Figure 3.1) and the TFCS NF element operated in the trailing vessel (referred to as the NF vessel). The DETU feed water (MFSE) was pumped to the vessels from a 100 gallon high-density polyethylene tank using a low- and high-pressure centrifugal pump at pressures ranging from 80-100 psi for unfouled elements. The feed flowrate and pressure to DETU was adjusted by the valve downstream of the high-pressure pump. The feed pressure was measured using an in-line pressure gauge. The test compounds were injected into the feed line using an electromagnetic dosing pump (Model no. A741-35257, LMI Milton Roy, Acton, MA, with a maximum dosing capacity of 0.58 gph) from 1 gallon dosing solution jars. The dosing pump was calibrated prior to the start of each test. Concentrate from the RO vessel flowed through a pressure control valve before entering the NF vessel. Permeate flowrate was regulated by two valves located in the vessel concentrate lines. Pressures were measured in the feed and concentrate flows of each vessel using a single pressure gauge coupled to a 4-way selector valve. Permeate flows and NF vessel concentrate flow were measured using in-line flowmeters (King Instrument, Huntington Beach, CA). Samples were collected from the feed (from an in-line sample port), two permeates, and the trailing-vessel concentrate. With the exception of the high-density polyethylene feed tank and permeate tubing, all system components in contact with water were made of 316 stainless steel in an effort to minimize corrosion and the introduction of corrosion byproducts into the membrane system. The DETU was operated in once-through mode, with no concentrate or permeate recycle.

The MSTU contained six vessels arranged in a two-stage 2:2:1:1 array. Each vessel was designed to hold three 4-inch diameter by 40-inch length elements (Figures 3.3 and 3.4). As with the DETU, the microfiltered secondary effluent was pumped through the vessel array from the feed tank using the low- and high-pressure centrifugal pumps common to both the DETU and MSTU in a single-pass mode. As with the DETU, feed pressure was measured using an in-line pressure gauge. Permeate flows were regulated by a valve located in the final concentrate line. Permeate and concentrate flows were measured using in-line flowmeters (No. 3\4-40-G-10, W&T). A single pressure gauge coupled to a 4-way selector valve was used to measure the concentrate differential pressures from vessels 1&2, 3&4, 5, and 6 (i.e., final concentrate) with respect to the feed pressure. Samples were collected from the feed, each vessel permeate, combined permeate, and final concentrate. Samples from the feed, final concentrate, individual vessel permeates and combined permeate were collected from in-line sample ports.

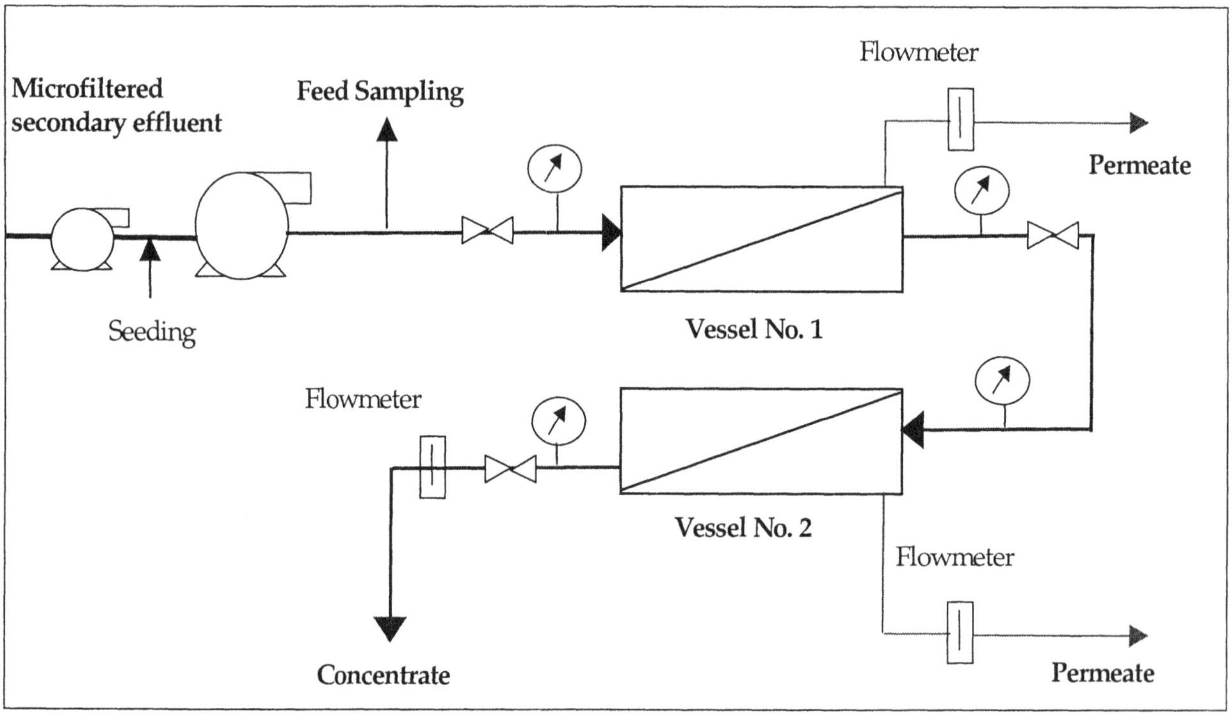

Figure 3.1 Schematic of the DETU

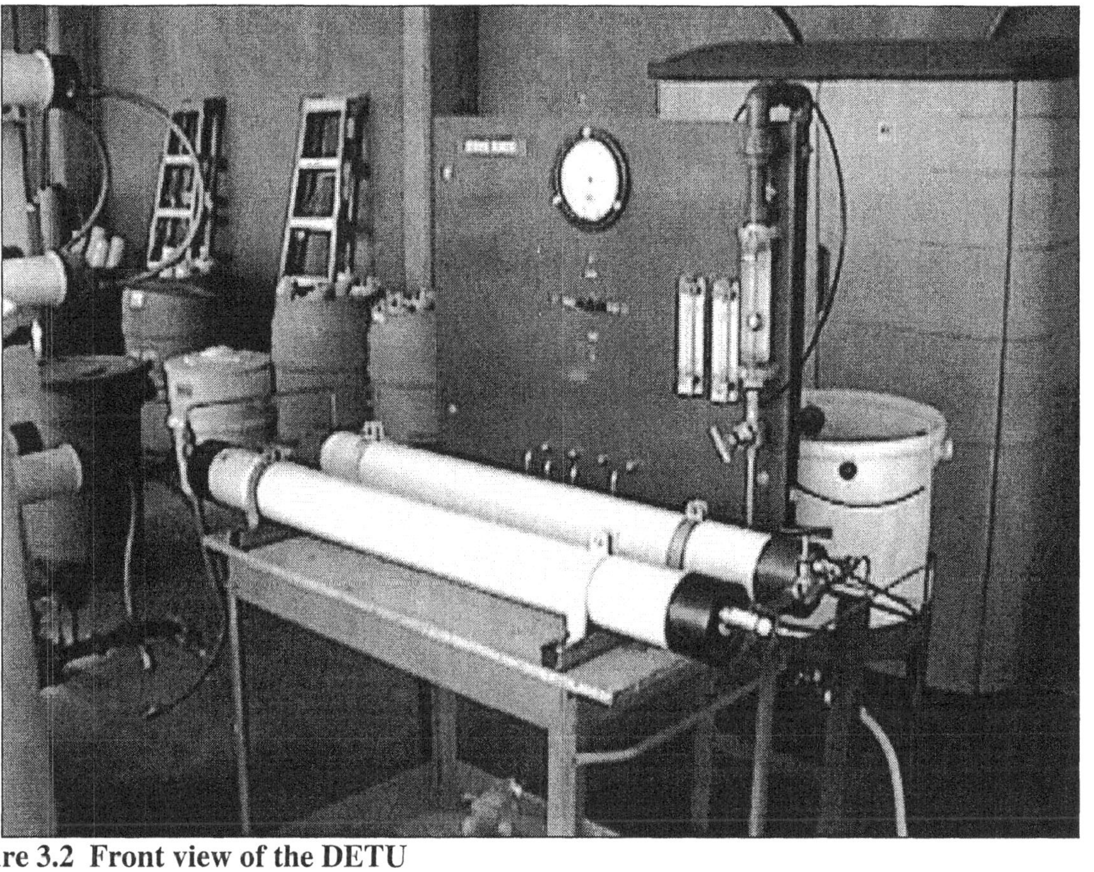

Figure 3.2 Front view of the DETU

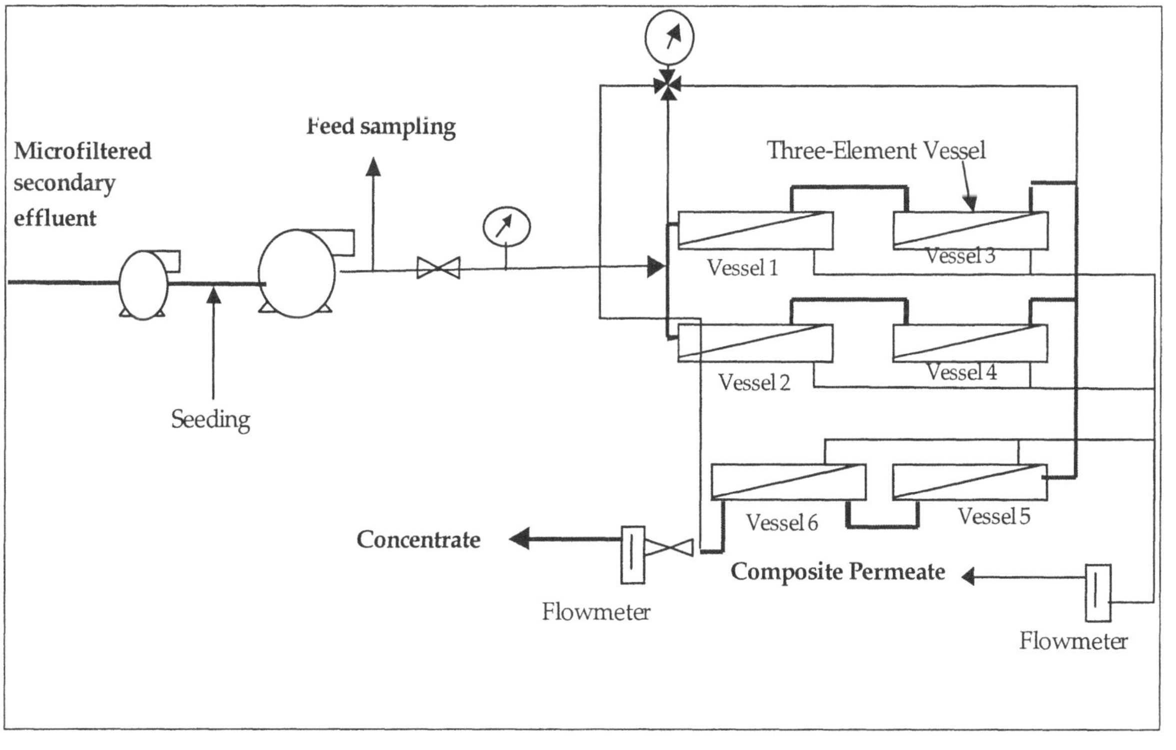

Figure 3.3 Schematic of the MSTU

Figure 3.4 Rear view of the MSTU

NWR-WTP.

Two units were employed for testing in NWR-WTP: a single-element test unit (SETU) (Figure 3.5) and a multi-vessel test unit (MVTU) (Figure 3.6). Figure 3.7 shows photo of the MVTU.

The SETU was designed to use a single 4-inch diameter by 40-inch long membrane element. The SETU was configured for batch or semi-batch operation using a feed/recirculation tank, but was operated in only batch mode during testing. A copper cooling coil was immersed into the feed solution tank to provide temperature regulation. In addition, a second copper cooling coil was installed on the concentrate return line that was immersed in a 55-gallon drum through which plant permeate water was recirculated to provide additional temperature regulation. Make-up water for the SETU consisted of permeate from the full scale surface water RO trains.

The MVTU contained three parallel pressure vessels, each capable of holding three 4" diameter by 40" long elements. It was configured so that feedwater flow, to which the surrogates were dosed, was directed to two of the three vessels, with each vessel containing three ESPA-1 RO or TFCS NF elements. This way, tests were conducted concurrently with both element types to reduce the amount of testing time. The MVTU was operated in a single-pass mode (no recirculation of either permeate or concentrate). Conventionally treated (coagulated, settled and filtered including 10-μm cartridge filters) river water (CTRW) was the feed to MVTU. MVTU feed water was extracted from a 4-inch PVC pipe existing in the pilot-room of NWR-WTP and then pumped to the vessels using a high-pressure centrifugal pump at pressures ranging from 60-70 psi. The flowrate and pressure to vessels were adjusted by the pressure control valves (Model EZ, Fisher Controls). Flowrates were measured using in-line flowmeters (Model no. 2-FE-2B 3M, King Instrument, Huntington Beach, CA). The test compounds were injected into the feed line using an electromagnetic dosing pump (Model no. P151-391SI, LMI Milton Roy, Acton, MA) from 1 gallon dosing solution jars. The dosing pump was calibrated prior to the start of each test. Samples were collected from the feed, two permeates and two concentrates using in-line sample ports.

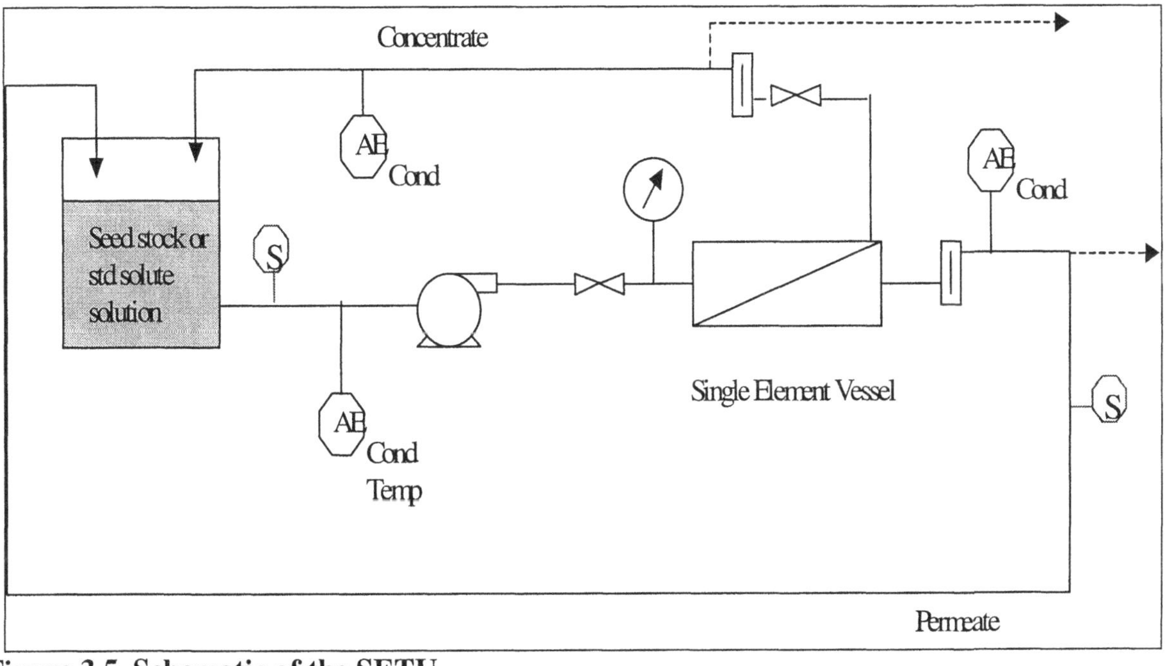

Figure 3.5 Schematic of the SETU

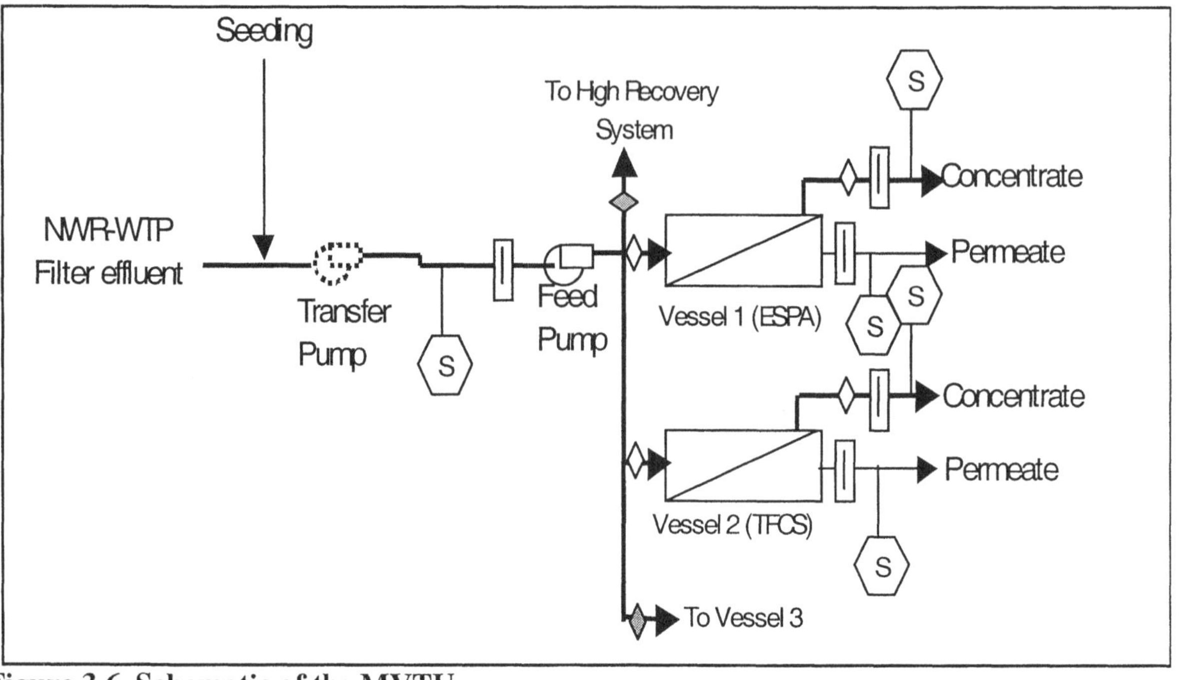

Figure 3.6 Schematic of the MVTU

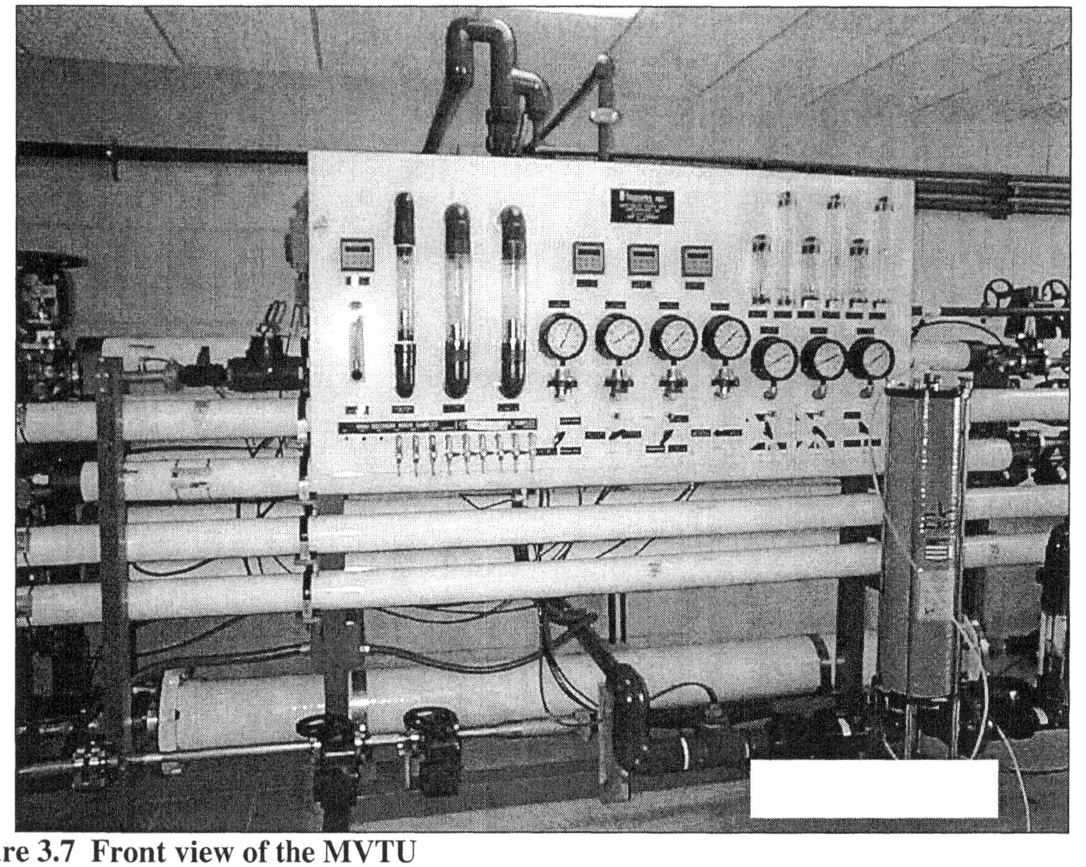

Figure 3.7 Front view of the MVTU

Table 3.1
General characteristics and performance specifications of the tested membranes

	Manufacturer	Hydranautics[a]	Koch Fluid Systems[b]
General Information	Commercial Name	ESPA1	TFCS
	Membrane Type	Reverse Osmosis	Nanofiltration
	Active Layer	Polyamide	Polyamide
	PH range	3–10	4-11
	Max. Press. (psi)	600	350
Performance Specifications	Solute/Conc. (mg/L)	NaCl/1500	NaCl/500
	Minimum Solute Rejection (%)	99 (for 15% recovery)	85[c] (for 10% recovery)
	Pressure (psi)	150	80
	Temperature (°C)	25	25
	PH range	6.5–7.0	6.5–7.0

[a] Oceanside, CA.
[b] San Diego, CA.
[c] measured as chloride ion.

Test Elements

Two spiral-wound elements, Hydranautics Model ESPA1-4040-UHT RO and Koch Fluid Systems Model TFCS-4040S NF, were tested in the pilot study. The characteristics and performance specifications of the tested membranes are tabulated in Table 3.1.

Analytical Techniques

The following sections describe the analytical techniques employed to measure the concentrations of the tested surrogates.

Rhodamine WT

R-WT concentrations in the samples were measured on-site using a TD-700 laboratory fluorometer (Turner Designs, Sunnyvale, CA). An excitation wavelength of 550 nm and an emission wavelength of 570-700 nm were used, together with a clear quartz lamp, to measure R-WT concentrations. The minimum quantification level for the NF/RO permeate was 20-25 ng/L. The R-WT calibration standards with concentrations ranging from 20 ng/L to 5 mg/L were prepared in DDW (obtained from the laboratories in OCWD and NWR-WTP) using 21.3% R-WT stock solution (Cat No. 10-108, Turner Designs, Sunnyvale, CA). Glass volumetric flasks with 100, 250, or 500 ml volumes were used in preparing the dilutions. New standard calibration curves were generated before each set of samples to be measured. Since the R-WT concentrations were significantly different (i.e., order of magnitudes) between the feed or concentrate (mg/L range) and permeate (ng/L-μg/L range) samples, and since the fluorometer exhibited much better sensitivity for relatively narrow standard concentration ranges, different calibration curves for different concentration ranges were generated. Figures B.1 and B.2 show example calibration curves for ng/L and mg/L concentration ranges, respectively. For all the calibration curves generated, r2 values ranged between 0.92-0.99.

Microsphere

A 2% solids (20,000 mg/L) stock solution of 0.02 μm orange fluorescent carboxylate-modified FluoSpheres® was obtained from Molecular Probes (Cat No. C-17261, Eugene, OR). The actual particle size was reported as 0.024 ± 0.005 μm. As with the bench-scale phase of this project, microsphere concentrations were determined using a method developed at the University of Illinois with a Bowman Series 2 Luminescence Spectrometer (SLM Aminco, Rochester, NY). An excitation wavelength of 533±4 nm or 533±8 nm (depending on concentration) and an emission wavelength of 566±8 nm were used to analyze all microsphere samples. Although the maximum emission/excitation wavelengths were 540/560 nm, the wavelengths of 533/566 nm were used to decrease interference due to light scattering. A high-quality quartz spectrofluorometry cuvette was used in order to minimize inconsistencies caused by cuvette imperfections. The minimum quantification level with this method was around 50 ng/L. Figures B.3 and B.4 represent typical calibration curves for ng/L and mg/L concentration ranges, respectively. A detailed discussion of the procedure used for microsphere concentration determination was presented in the previous section for the bench-scale portion of this report.

MS2 Phage

For the determination of MS2 *phage* concentrations in the samples, a plaque assay procedure that was originally developed by the EPA's Cincinnati laboratories was employed at the University of Illinois. The plaque assay procedure involves growing a lawn of *E. coli* in the presence of the MS2 *phage* sample to be measured. The MS2 *phage* viruses inhibit the growth of *E. coli* cells in near proximity, thereby producing plaques (points of no *E. coli* growth) in the *E. coli* lawn. Given proper dilution, each plaque can be counted and the concentration of MS2 *phage* in the original sample can be determined. Again a detailed discussion of the procedure used for MS2 *phage* concentration determination was presented previously.

Sodium Chloride

For the SETU experiments conducted in NWR-WTP, sodium chloride concentration measurements were performed using an Orion Model 135 conductivity meter, having a minimum detection limit of 0.1 μS/cm. Conductivities of 500 and 1,500 mg/L NaCl solution used during standardizations were approximately 1,000 and 3,000 μS/cm, respectively. At these feed conductivities and the degree of NaCl rejection achieved by the RO and NF membranes, the conductivity meter can measure a maximum of 4-log rejection of the RO membrane and < 2-log rejection for the NF membrane based on the minimum detection limit. Conductivity of the CTRW at the NWR-WTP ranged between 250 and 2,500 μS/cm during SETU and MVTU tests. At this element feed concentration and the minimum detection level of the meter, the maximum removal measurable across the membrane was about 3.5 log, again greater than the rated log removal of conductivity by either membrane tested. A standard calibration curve of conductivity versus sodium chloride concentration is shown in Figure B.5.

Conductivity, TDS, pH, and Temperature

For the DETU, MSTU and MVTU testing, conductivity, total dissolved solids (TDS), pH, and temperature of the samples were measured onsite during the conduct of each experiment using a single meter that had separate probes for conductivity/TDS and pH (Ultrameter 6P, Myron L company, Carlsbad, CA). The meter was calibrated for pH using a pH 7 standard buffer solution (Cat No. 910107, Orion), and for conductivity/TDS using a 0.01 M KCl solution (Cat No. 5890-32, Ricca Chemical).

EXPERIMENTAL APPROACH

Experimental approach employed at two sites is summarized in the following subsections.

WF-21

The experimental pilot test matrix employed in WF-21 is shown in Table 3.2. The elements tested were Hydranautics RO Model ESPA1-4040-UHT and Koch Fluid Systems NF Model TFCS-4040S. The first phase DETU experiments employed elements in the following matrix of membrane and o-ring conditions:

- Intact membrane and intact o-rings
- Compromised elements and intact o-rings
- Intact membrane and compromised o-rings

In addition, the impact of the fouling and subsequent chemical cleaning of membrane-compromised elements on removals of the surrogate indicators was investigated in this phase. Target flux and recovery in each DETU experiment was 10.4 gfd and 15%, respectively. The low flux was selected as representative of full-scale RO/NF element operation on MFSE.

The second phase MSTU experiments (labeled W 6 to W 13 in Table 3.2) included testing of a single compromised ESPA1 RO elements or o–ring installed at different locations within the vessel array of the MSTU. For the compromised membrane tests, the 17 remaining positions of the vessels in MSTU were installed with uncompromised RO elements. Figure 3.8 shows the schematic of the MSTU with respect to array and vessel positions. Target flux and target recovery in each MSTU experiment was 10.4 gfd and 75%, respectively. For all the DETU and MSTU experiments, MFSE was used as the feed. Table 3.3 shows typical concentration values for some of the physiochemical and microbiological characteristics of the MFSE. Figure 3.9 shows a picture of the demonstration-scale US Filter MEMCOR CMF-S MF unit that provides the feedwater for the SETU and MVTU tests.

Each experiment included the dosing of microspheres, MS2 *phage* and R-WT in that order. The R-WT was dosed last to avoid the positive interference of the fluorescent dye molecules on the measurement of the fluorescent microspheres, particularly in the permeates. The dosing solutions of microsphere, MS2 and R-WT were prepared in DDW. Each set of experiments is discussed in further detail later in this section.

Table 3.2
Experimental matrix for pilot-scale tests conducted at WF-21

Test	Unit[a]	Test Type	Position in Array	Water Source[b]	Membrane Type	Membrane Condition	Target Microsphere Conc (mg/L)[c]	Target MS2 Conc (pfu/ml)[d]	Target R-WT Conc (mg/L)[e]	Target Flux (gfd)	Target Recovery (% of feed)
W-1a	DETU	Uncompromised[f]	NA	MFSE	ESPA1&TFCS[i]	new	1			10.4	15
W-1b	DETU	Uncompromised	NA	MFSE	ESPA1&TFCS	new		1.0E+07		10.4	15
W-1c	DETU	Uncompromised	NA	MFSE	ESPA1&TFCS	new			1	10.4	15
W-2a	DETU	compromised-M[g]	NA	MFSE	ESPA1&TFCS	new	1			10.4	15
W-2b	DETU	compromised-M	NA	MFSE	ESPA1&TFCS	new		1.0E+07		10.4	15
W-2c	DETU	compromised-M	NA	MFSE	ESPA1&TFCS	new			1	10.4	15
W-3a	DETU	compromised-O[h]	NA	MFSE	ESPA1&TFCS	W-1a	1			10.4	15
W-3b	DETU	compromised-O	NA	MFSE	ESPA1&TFCS	W-1b		1.0E+07		10.4	15
W-3c	DETU	compromised-O	NA	MFSE	ESPA1&TFCS	W-1c			1	10.4	15
W-4a	DETU	compromised-M	NA	MFSE	ESPA1&TFCS	W-2a/fouled	1			10.4	15
W-4b	DETU	compromised-M	NA	MFSE	ESPA1&TFCS	W-2b/fouled		1.0E+07		10.4	15
W-4c	DETU	compromised-M	NA	MFSE	ESPA1&TFCS	W-2c/fouled			1	10.4	15
W-5a	DETU	compromised-M	NA	MFSE	ESPA1&TFCS	W-2a/fouled/cleaned	1			10.4	15
W-5b	DETU	compromised-M	NA	MFSE	ESPA1&TFCS	W-2b/fouled/cleaned		1.0E+07		10.4	15
W-5c	DETU	compromised-M	NA	MFSE	ESPA1&TFCS	W-2c/fouled/cleaned			1	10.4	15
W-6	MSTU	compromised-M	lead vessel (vessel no.1)	MFSE	ESPA1	new	1	1.0E+07	1	10.4	75
W-7	MSTU	compromised-M	vessel no.3	MFSE	ESPA1	W-6	1	1.0E+07	1	10.4	75
W-8	MSTU	compromised-M	vessel no.5	MFSE	ESPA1	W-6	1	1.0E+07	1	10.4	75
W-9	MSTU	compromised-M	trailing vessel (vessel no.6)	MFSE	ESPA1	W-6	1	1.0E+07	1	10.4	75
W-10	MSTU	compromised-O	lead vessel (vessel no.1)	MFSE	ESPA1	new	1	1.0E+07	1	10.4	75
W-11	MSTU	compromised-O	vessel no.3	MFSE	ESPA1	W-10	1	1.0E+07	1	10.4	75
W-12	MSTU	compromised-O	vessel no.5	MFSE	ESPA1	W-10	1	1.0E+07	1	10.4	75
W-13	MSTU	compromised-O	trailing vessel (vessel no.6)	MFSE	ESPA1	W-10	1	1.0E+07	1	10.4	75

[a] DETU: dual element test unit; MSTU: multi-stage test unit.
[b] MFSE: microfiltered secondary effluent.
[c] Target concentrations based on bench testing.
[d] pfu: plaque forming unit.
[e] R-WT: Rhodamine WT.
[f] Standardization tests used both new and compromised membranes to identify extent of imperfection required to achieve desired passage.
[g] Compromised membrane.
[h] Compromised o-ring.
[i] For all the DETU experiments, ESPA1 and TFCS elements were always in the first and second vessels, respectively.
Note: Microspheres were injected before R-WT in each experiment to prevent potential flourescence measurement interference.
NA: not applicable.

Table 3.3
Typical characteristics of the microfiltered secondary effluent (MFSE) at WF-21 as feed to DETU and MSTU

BOD_5 (mg/L)	Total Nitrogen (mg/L)	Phosphate (mg/L, PO_4-P)	Temperature (°C)	pH	MS2 *phage* (pfu/ml)	Conductivity (μS)	TDS (mg/L)
10-15	10-20	1.5-3	24-27	7.4-7.9	0[a]	1,450-1,650	1,000-1,150
Sodium (mg/L)	**Calcium (mg/L)**	**Magnesium (mg/L)**	**Chloride (mg/L)**	**Sulfate (mg/L)**	**Alkalinity (mg/Las $CaCO_3$)**		
200-215	70-85	25-30	200-220	290-310	200-250		

[a] Zero concentration was found for all the six samples analyzed. Combined chlorine in the samples was quenched immediately with excess amount of sodium thiosulfate.

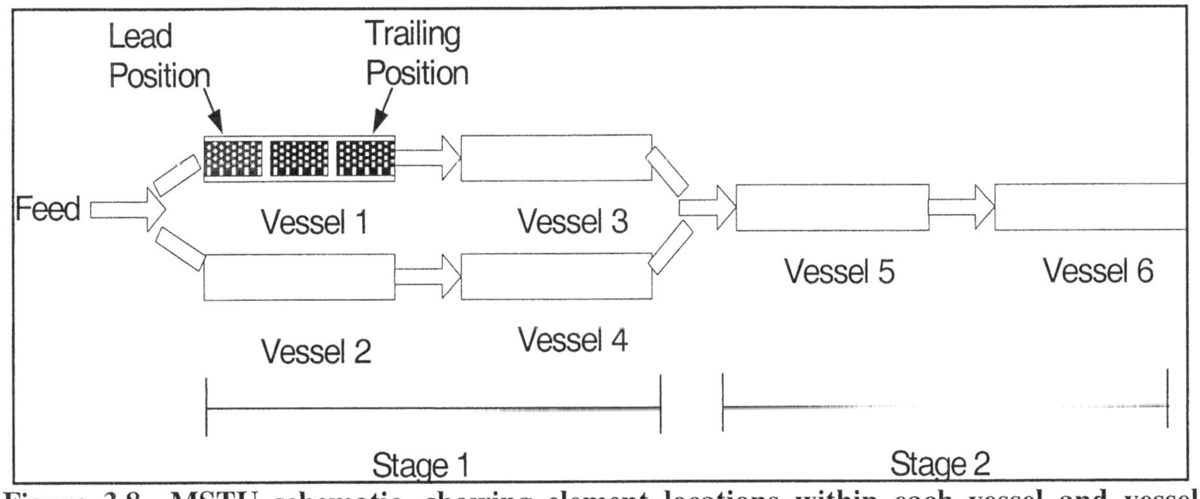

Figure 3.8 MSTU schematic, showing element locations within each vessel and vessel positions in array

Figure 3.9 A picture of the demonstration-scale U.S. Filter MEMCOR CMF-S MF unit in operation at OCWD

NWR-WTP

The experimental pilot test matrix employed at the NWR-WTP is shown in Table 3.4. The elements tested were identical to those used at WF-21: Hydranautics RO Model ESPA1-4040-UHT RO and Koch Fluid Systems NF Model TFCS-4040S. The first phase SETU experiments employed elements having intact membrane and o-rings for standardization experiments. Make-up water for preparation of solute (sodium chloride) stock solutions and surrogates was permeate from the NWR-WTP RO system (plant permeate). Target flux was 12 gfd typical of RO/NF flux on conventionally treated surface waters. Target recovery in each SETU experiment was 15% and 10% for RO and NF elements, respectively as recommended by the membrane suppliers.

Similar to DETU experiments in WF-21, the second phase MVTU experiments employed RO and NF elements and o-rings in the following matrix:

- Intact membrane and intact o-rings
- Compromised elements and intact o-rings
- Intact membrane and compromised o-rings

The MVTU was configured so that feedwater flow, to which the surrogates were dosed, was directed to two of the three vessels, with each vessel containing three RO or NF elements.

Thus, tests were conducted concurrently with both element types reducing the amount of testing time and the mass of surrogates. The impact of fouling by particulates and dissolved organics in the CTRW and subsequent chemical cleaning of membrane-compromised elements on removals of the surrogates was also investigated. Furthermore, similar to MSTU experiments, the impact of the location of the compromised membrane or o–ring on removals was also investigated. Therefore, the impacts of these factors on the surrogate removals were compared for a microfiltered secondary wastewater (as tested in WF-21) and a conventionally treated and filtered surface water. As with the tests in WF-21, each experiment included the dosing of microspheres, MS2 *phage* and R-WT. R-WT was dosed last to avoid the positive interference of the fluorescent dye molecules on the measurement of the fluorescent microspheres, particularly in the permeates. The dosing solutions of microsphere, MS2 and R-WT were prepared in DDW obtained from the laboratory at the NWR-WTP.

Target flux and target recovery in each MVTU experiment was 12 gfd and 45%, respectively. For all the MVTU experiments, CTRW polished by 10-μm cartridge filtration served as the feed. Table 3.5 shows typical concentration values for some of the physiochemical and microbiological characteristics of the CTRW.

Table 3.4
Experimental matrix for pilot-scale tests conducted at NWR-WTP

Test	Unit[a]	Test Type	Position in Array	Water Source	Membrane Type	Membrane Condition	Target Microsphere Conc (mg/L)[g]	Target MS2 Conc (pfu/ml)[g]	Target R-WT Conc (mg/L)	Target Flux (gfd)	Target Recovery (% of feed)
RO-S-1	SETU	flux/rejection standardization[b]	NA	RO permeate/ 1500 mg/L NaCl	ESPA1	new				100 psig	15
RO-S-2	SETU	R-WT standardization	NA	1500 mg/L NaCl	ESPA1	RO-S-1			1	12	15
RO-S-3	SETU	MS2 standardization	NA	1500 mg/L NaCl	ESPA1	new		1.0E+07		12	15
RO-S-4	SETU	Microsphere standardization	NA	1500 mg/L NaCl	ESPA1	new	1			12	15
NF-S-1	SETU	flux/rejection standardization[c]	NA	RO permeate/ 500 mg/L NaCl	TFCS	new				80 psig[c]	10
NF-S-2	SETU	R-WT standardization	NA	500 mg/L NaCl	TFCS	new			1	12	10
NF-S-3	SETU	MS2 standardization	NA	500 mg/L NaCl	TFCS	new		1.0E+07		12	10
NF-S-4	SETU	Microsphere standardization	NA	500 mg/L NaCl	TFCS	new	1			12	10
C-1	MVTU	uncompromised	NA	CTRW[f]	ESPA1&TFCS	new	1	1.0E+07	1	12	45
C-2	MVTU	compromised-M[d]	lead element	CTRW	ESPA1&TFCS	C-1	1	1.0E+07	1	12	45
C-3	MVTU	compromised-M	trailing element	CTRW	ESPA1&TFCS	C-1	1	1.0E+07	1	12	45
C-4	MVTU	compromised-M	lead element	CTRW	ESPA1&TFCS	C-1/fouled	1	1.0E+07	1	12	45
C-5	MVTU	compromised-M	lead element	CTRW	ESPA1&TFCS	C-1/fouled/cleaned	1	1.0E+07	1	12	45
C-6	MVTU	compromised-O[e]	lead element	CTRW	ESPA1&TFCS	new	1	1.0E+07	1	12	45
C-7	MVTU	compromised-O	trailing element	CTRW	ESPA1&TFCS	new	1	1.0E+07	1	12	45

[a] SETU: single element test unit; MVTU: multi-vessel test unit
[b] Element was operated at Hydranautics standard test condition feed pressure of 150 psig, 15% recovery, pH 6.5-7.0 and as close to 25 °C as possible.
[c] Element was operated at Fluid Systems standard test condition feed pressure of 80 psig, 10% recovery, pH 7.0 and as close to 25 °C as possible.
[d] Compromised membrane.
[e] Compromised o-ring.
[f] Coagulated, settled, filtered (including 10-μm filtration) surface water.
[g] Target concentrations based on bench testing.
Note: Microspheres were injected before R-WT in each experiment to prevent potential flourescence measurement interference.
NA: not applicable.

Table 3.5
Typical characteristics of the CTRW at NWR-WTP as feed to MVTU

Parameter	Typical range
Ca (mg/L)	11-17
Mg (mg/L)	3-7
Na (mg/L)	37-74
K (mg/L)	2-6
Fe (mg/L)	0.005-0.09
Mn (mg/L)	0.009-0.1
NO_3+NO_2-N (mg/L N)	0.2-0.45
NH_4-N (mg/L N)	0.05-0.1
Cl (mg/L)	30-300
SiO_2 (mg/L)	2-10
SO_4 (mg/L)	50-110
HCO_3 (calc. From Alk.)	10-37
Br (mg/L)	ND[a]-2.4
PO_4-P (mg/L)	ND-0.2
Conductivity (μS)	750-2,500
Alkalinity (mg/L $CaCO_3$)	8-30
Hardness (mg/L $CaCO_3$)	35-100
SDI	3-4
Turbidity (NTU)	0.08-0.15
TDS (evap. @180 °C) (mg/L)	150-600
Temperature (°C)	20-27
PH	5.9-6.4
TOC (mg/L)	8-14
UV_{254} (1/cm)	0.14-0.23
$SUVA_{254}$ (L/mg-m)	1-2.8
Coliforms (#/100 ml)	<1-10
HPC (cfu/ml)	<1-500

[a] ND: not detected.

Surrogate Testing Procedures

The following subsections summarize the detail testing procedures for all experiments conducted. Since the DETU and MSTU at WF-21 and MVTU test at NWR-WTP had very similar surrogate testing procedures, they are discussed together under the same subsection. Some minor differences in protocols between the two sites are noted throughout the text, however. In contrast, the protocols for the first phase experiments at the NWR-WTP (SETU tests) are discussed separately as the experiments had different experimental objectives and experimental approach.

SETU Experiments

Solute Standardization. All RO and NF elements to be used in NWR-WTP experiments were first tested in the SETU on sodium chloride (NaCl) feed solution. NaCl solution was prepared by adding crystalline NaCl to a given amount of plant permeate to achieve the desired conductivity. The booster pump was then used to recirculate the permeate in the feed tank, dissolve the NaCl, and provide a uniform concentration of solute prior to operation of the SETU (Figure 3.5). Once the challenge solution was prepared, solute testing began by starting the feed pump on the SETU. Feed, permeate, and concentrate samples were collected after 30, 45, and 60 minutes of operation and tested for conductivity and temperature. In addition, system pressures and flows were measured and recorded during each sampling event. Data gathered during solute standardization to demonstrate the integrity of each element by confirming salt rejection was compliant with the manufacturer's specification.

R-WT Standardization. R-WT testing was performed in the SETU to determine the level of R-WT rejection by both element types and confirm analytical capability to measure R-WT in the permeate with the field fluorometer. One of the RO or NF elements tested in the previous NaCl test was used. A target feed concentration of 1 mg/L was chosen based on the advertised capability to detect 10 ng/L R-WT in permeate. Detection at such low levels would provide a maximum sensitivity of 5-log removal through the spiral-wound element based on the 1 mg/L feed concentration. In addition, background solute concentrations of 1,500 mg/L NaCl for Hydranautics membranes and 500 mg/L for Koch membranes were used to determine potential membrane fouling by R-WT. A reduction in flux or a change in NaCl rejection following R-WT addition would indicate the R-WT dye was causing membrane fouling.

The R-WT challenge solution was prepared by adding a measured volume of R-WT stock solution to a given volume of plant permeate to achieve the desired R-WT concentration. The booster pump was then operated to fully mix the challenge solution and provide a uniform concentration of R-WT prior to commencement of testing. Once the feed tank was adequately mixed (as judged from collection and measurement of feed samples at various times prior to testing), testing began by starting the feed pump on the SETU. Feed, concentrate, and permeate samples were collected on 15-minute intervals for 2 hours. Samples were measured for fluorescence, conductivity, and temperature. In addition, the flow and pressure of each flow stream was recorded during sampling events.

MS2 Phage Standardization. Once solute and R-WT testing were completed for RO and NF element, another new element (both RO and NF) tested for NaCl rejection initially was installed in the SETU and tested for MS2 *phage* rejection. Target feed concentrations of 10^7 pfu/mL were not achieved, however, due to weaker than expected stock *phage* solutions. Feed concentrations were 2 logs less than anticipated (10^5 pfu/mL) even though strict quality control measures were followed during preparation, shipping, and handling of virus stock and collected samples. The inability to attain target *phage* concentrations was subsequently attributed to a defective strain of *phage* as provided by ATCC.

To conduct the virus experiments, the feed tank was first filled with a minimum volume of plant permeate and NaCl was added to achieve the target solute concentration. Next, virus stock solution was added. Once the *phage* stock was added to feed tank, the booster pump was started to recirculate the feed solution and attain a uniform concentration. After the phage was adequately mixed in feed tank, the valve to the test stand pump was opened and testing begun. Feed samples were collected at target times of 0, 1.5, 3, 6, 9, 12, 15, 20, 30, 45, and 60 minutes. These samples were used to assess virus uptake by the system. Feed samples were collected from the feed tank using disposable pipettes to minimize the risk of contamination of laboratory equipment and permeate samples. Once the feed samples were collected and deposited into sample vial, the pipette was discarded and the sample vial was cleaned using an isopropyl alcohol solution. The collected samples were then shipped overnight to the University of Illinois for analysis.

Permeate samples were collected from the permeate sample port located on the SETU. The permeate valve was opened and allowed to drain for 1 minute prior to collection of permeate sample. This ensured the sample was representative of the permeate being produced by the element at any given time. Permeate samples were collected at 10, 20, 45, and 60 minutes after the addition of virus stock to the feed solution.

Once all feed and permeate samples were collected feedwater was drained from the system and the tank refilled with plant permeate. While still allowing permeate to flow into the feed tank, both booster and feed pumps were energized and the concentrate directed to drain to remove as much virus solution from the SETU as possible. The SETU was then flushed with approximately 100 gallon of plant permeate, after which a 50 mg/L sodium hypochlorite (NaOCl) solution was batched in the SETU feed tank and circulated through the system for 40 minutes to inactivate remaining *phage*. Sodium bisulfite (SBS) was then added to the feed tank at greater than stoichiometric concentrations to the chlorine residual and circulated through the system for 15 minutes to neutralize the residual NaOCl. The system was then flushed with plant permeate to remove the SBS. The final chlorine concentration in the system was verified to be non-detect. After flushing of the system, the element was removed and stored in 0.5% SBS preservative solution.

Standardization Testing of Microsphere for Fouling and Decay Determination. Tests were conducted with one of NF and RO elements in the SETU to quantify (1) membrane fouling caused by the accumulation of microspheres at the surface of the membrane and (2) the extent to which microspheres were adsorbed on surfaces of the SETU, thereby, decreasing their concentration in the membrane feed. Similar to R-WT and virus tests, NaCl was added to the microsphere test solution. Concentration of microspheres in the feed tank were selected to account for projected loss (adsorption) of these particles on the surfaces of the SETU. The degree of loss was based on uptake data measured during bench-scale testing. Prior to their introduction into the feed water, the 0.02 μm microsphere stock solution was sonicated for 5 minutes to separate any clusters of microspheres, which may have formed during storage. Microsphere stock solution was added to the feed tank approximately 10 minutes before collection of permeate and feed samples into approximately 150-gallons of feedwater to achieve a feed concentration of 1 mg/L. This volume was chosen to minimize the mass of microspheres required (thereby, controlling cost), while still providing sufficient mass for microsphere challenge testing in the SETU.

The booster pump was then operated in recirculation mode to thoroughly mix the microspheres. Feed samples were collected periodically and tested for fluorescence using the field fluorometer to ensure adequate mixing. Once consistent fluorescent measurements were observed with the fluorometer and the 1 mg/L feed dose was verified, the feed decay experiment was initiated.

Each experiment was initiated once the test stand pump was activated. The system was operated at identical conditions used for virus experiments. Feed samples were collected at 0, 1.5, 3, 6, 9, 12, 15, 20, 30, 45, and 60 minutes from the feed sample tap on the SETU. Two 40-mL vials of feed sample were collected for each sampling event. One vial was used to "field verify" feed concentrations and the other vial was shipped to the University of Illinois to be tested in the bench top spectrofluorometer to provide the actual concentration. The field fluorometer used to measure R-WT was also used to make "rough" checks of the feed concentration of microspheres.

During each experiment, a permeate sample was collected between 5 and 10 minutes after microsphere seeding, which was used to assess microsphere passage. The permeate sample valve was opened 5 minutes before collection of the sample to allow permeate to drain into a "catch container." This ensured the permeate line was fully flushed for the collection of each permeate sample. Before the collection of each additional permeate and feed sample, the feed concentration of the microspheres was measured using the field fluorometer. If the feed concentration was less than 0.5 mg/L, an additional volume of sonicated microsphere stock was added to the tank to bring the feed concentration of microspheres to a target value of 1 mg/L. Microspheres were seeded 5-10 minutes before each sampling event as necessary.

After all feed and permeate samples were collected the test stand pump was turned off, the microsphere/NaCl solution drained from the system and the feed tank filled with plant permeate. Approximately 10 mL of Tween 20 concentrate (surfactant) was added to the feed tank and the booster pump operated in recirculation mode for 30 minutes to mix Tween 20 and scour microspheres from feed tank surfaces. The feed tank was then emptied and the procedure repeated. The feed tank was then filled with plant permeate and 10 mL of Tween 20 and the feed pump activated to send the concentrate and permeate to drain. This served to remove microspheres from the SETU and prevent contamination when elements were changed. Finally, the system was flushed with a full tank of new plant permeate to remove Tween 20 from the system. After flushing, the element was removed and preserved.

DETU, MSTU and MVTU Experiments

Microsphere Testing. Microsphere was the first surrogate injected to the system for each experimental set. In order to break down potential clusters of microspheres formed during storage and homogenize the solution, the 2% stock solutions were first sonicated for 5 minutes before their use for the dosing solution preparation. The concentration of the dosing solution was selected based on the feed and dosing solution flowrates to achieve the target microsphere concentration of 1 mg/L to the elements. After all the dosing and cleaning solutions were prepared, the membrane feed pump for a test unit was activated, and the flows and pressures adjusted to target conditions. The test units were operated for 60 minutes, during which time pressures and flows were recorded; and the conductivity, TDS, pH, and temperature of the feed, permeates and concentrate were measured every 15 minutes to confirm each element was operating within advertised performance requirements. After the 60-min of initial running, samples from feed and permeates were collected to be sent to University of Illinois to establish the background fluorescence readings for the microsphere analysis. The dosing pump was calibrated for the delivery of the required flow, and the flowrate was double-checked again with pumping DDW right before the microsphere dosing started.

In the initial experiments, 5 min (for DETU), 7 min (for MSTU) and 8 min (for MVTU) of dosing was found to be sufficient time for microspheres to uniformly mix and disperse from the dosing point to the feed sample point, and to achieve a steady feed concentration of approximately 1 mg/L, which could be estimated by field analysis using the fluorometer. The field fluorometer used to measure R-WT was also used to make "rough" checks of the feed concentration of microspheres. Although not highly accurate, such measurements were of benefit in confirming the presence and concentration of microspheres within 20% of actual level. After these initial findings, all experiments incorporated the necessary initial dosing of 5-8 minute period to reach steady-state feed level of any test compound before the start of the actual experiment (i.e., t=0). Samples from feed, permeates and concentrates were collected at 5, 10, 15, and 30 min for DETU and at 5, 15, and 25 min for MSTU and MVTU during surrogate dosing. Sampling time for MSTU and MVTU was 25 min due to the larger feed flowrates and thus larger mass of costly microspheres required to achieve the challenge concentration. The samples were shipped to University of Illinois (along with MS2 *phage* samples) via over-night courier in coolers packed with synthetic ice. Pressures, flows, conductivity, TDS, pH, and temperature of feed, permeates and concentrates were also recorded in conjunction with collection of each set of samples. Feed and concentrate samples were collected, packaged and shipped separately from the permeate samples to prevent potential cross-contamination of the permeate samples from higher-concentration feed/concentrate samples.

After 25 or 30 min dosing, the dosing pump was turned off and the dosing tubing was disconnected from the injection fitting. The test units were then operated for approximately 60 minutes to flush out residual microspheres prior to the start of MS2 dosing. During the flushing period, pressures, flows, conductivity, TDS, pH, and temperature were recorded every 30 minutes to confirm element performance characteristics. The suction and discharge tubing of the dosing pump were disconnected from the pump and cleaned thoroughly with tap water from a hose connection. The dosing system was then re-assembled and cleaned by pumping a solution of Tween 20 (1 ml/L) through the tubing and pump for a period of 20 minutes. The Tween 20 solution was prepared in DDW. Residual Tween 20 solution was removed from the system by first rinsing the tubing with tap water using the hose connection and then pumping DDW through the tubing and pump for an additional 20-minute period. The discharge tubing was then reconnected to the injection fitting for MS2 dosing.

MS2 Phage Testing. The MS2 *phage* dosing solution was prepared in DDW using the MS2 stock provided by the University of Illinois. A new stock was used for each experiment, except when two experiments were conducted within 2-3 days, in which case a common stock was used. Stocks typically contained 10^{11}-10^{12} pfu/ml of virus. Viral concentration of the dosing solution was based on feed and dosing solution flowrates to provide a *phage* concentration in the element feed of approximately 10^7 pfu/ml.

For the DETU and MSTU experiments in WF-21, the microfiltered secondary effluent (MFSE) as the feed to the test units contained approximately 3-5 mg/L of combined chlorine, primarily in the form of chloramines, to minimize bacterial growth within the microfiltration and downstream RO system demonstration units. Free chlorine was not present in the MFSE due to the high concentration of ammonia-nitrogen (typically 10-20 mg/L). This was confirmed periodically throughout the experiments by OCWD personnel. Chloramines are weak oxidants and have the potential to cause damage to MS2 *phage*, decreasing their viability. To prevent damage, the MFSE was dosed with sodium thiosulfate to neutralize the combined chlorine at a dose in excess of the stoichiometric amount. Combined chlorine measurements were made before each experiment to confirm the absence of a detectable residual. For the MVTU experiments in NWR-WTP, the feed to the test unit did not contain any free or combined chlorine. However, the operators were periodically (twice a month) flushing the lines with chlorine. Experiments were not conducted during these periods. In an effort to be more conservative, considering the possibility of some residual chlorine in the lines after flushing, sodium thiosulfate was added to the feed for each experiment.

Before each *phage* experiment, flowrates of the dosing and feed pumps were confirmed. All the sampling ports were disinfected by wiping with Kim-Wipes soaked in 10% sodium hypochlorite. Feed/concentrate and permeate samples were handled and shipped separately to prevent cross-contamination. All the sampling vials used were sterile as purchased from the supplier.

As with the microspheres, the *phage* solution was dosed to the feed for an initial period of 5-8 min depending on the test unit to achieve steady-state concentration prior to start of each experiment. Samples of feed, permeates and concentrates were then collected at 5, 15, 30, and 45 minutes and shipped to University of Illinois by over-night courier in coolers packed with synthetic ice. Pressures, flows, conductivity, TDS, pH, and temperature were also measured during each sample collection.

At the completion of each test, the dosing pump was turned off, and discharge tubing was disconnected from the injection fitting. The system continued to operate 60 minutes until the start of the R-WT dosing to displace the dosed material. Pressures, flows, conductivity, TDS, pH, and temperature were also measured every 30 min during flushing. The dosing system was cleaned using the same procedure described in the previous "Microsphere Testing" section, except that a 50 mg/L hypochlorite solution was used instead of Tween 20. The cleaned system was then re-connected to the injection fitting for use with the R-WT testing.

R-WT Testing. R-WT dosing solution was prepared in DDW from the 21.3% stock solution. The target R-WT feed concentration was 1 mg/L. As with the previous tests, the dosing flowrates were double-checked right before the beginning of the R-WT experiment. After the initial dosing to achieve constant feed concentration, the experiment was started, and feed, permeate and concentrate samples were collected at 5, 15, 30, and 45 minutes, in addition to the measurement of pressures, flows, conductivity, TDS, pH, and temperature. Samples were collected in 4 oz. amber sample bottles.

At the completion of each test, the dosing pump was turned off, and discharge tubing was disconnected from the injection fitting. The units were operated for an additional 60 minutes to collect final data on pressures, flows, conductivity, TDS, pH, and temperature as well as to flush residual R-WT from the test units. Data were collected every 30 min during flushing.

It was found in the initial experiments that R-WT is taken up by the membranes, especially by NF, during dosing and then released in slowly into the permeate. If not flushed adequately, the background fluorescent R-WT dye would cause a positive interference with the analysis of microsphere when this surrogate was used in the subsequent experiment. In pre-experiment testing, it was determined that flushing of the test units for a period of 24 hours (at a feed flowrate of 4.3 and 10.5 gpm for DETU and MVTU, respectively) was sufficient to reduce permeate R-WT concentrations to detection levels. Therefore, for all the experiments, test units were flushed with feed water for 24 hours after the termination of R-WT dosing. Detailed discussion regarding R-WT uptake by the membranes is discussed later in a separate section.

After the completion of flushing for R-WT, and where elements were not to be used in the next experiment, the RO and NF elements were removed and stored in 0.5% sodium bisulfite preservative solution. The dosing system was cleaned using the same procedure described for microsphere testing, except that much more vigorous rinsing with a high-velocity spray was necessary to remove the more adsorptive dye from the surfaces of pump and tubings. In addition, the dosing system was flushed by pumping DDW for a period of 1-2 hours until the concentration of R-WT was decreased to background levels.

Membrane Fouling and Cleaning

In order to investigate the impact of membrane fouling and subsequent cleaning on the removals of surrogates by compromised membranes, compromised ESPA1 RO and TFCS NF elements were fouled in both the DETU and MVTU by operating both element types for prolonged periods (up to 2 weeks of continuous operation) with feedwaters. For DETU, the elements were operated at 15.8 gfd with 15% recovery until the feed pressure increased to 1.3 times the initial pressure at constant permeate flux. The value of 1.3 was chosen as the criteria that would represent a typical condition in a typical full-scale treatment at which membrane elements would be subject to a chemical cleaning. It took 10 days of continuous running to foul the membranes to this level in the DETU using MFSE. In contrast, operation for a period of 13 days at flux rates as high as 16 gfd resulted in a feed pressure increase of 1.15 times initial pressure.

Once the membranes were fouled, the experiments to investigate the impact of fouling on the surrogate removals were immediately conducted.

After completion of the compromised/fouled membrane experiments, the fouled elements were immediately cleaned using the cleaning skids available in WF-21 or NWR-WTP. At WF-21, 100 gallons of dechlorinated tap water was collected in a 150-gallon polyethylene tank and heated to 104°F using a heating coil in the tank. Sodium tripolyphosphate (16.7 lb) and sodium dodecylbenzene sulfonate (1.7 lb) was added to the heated water. The chemicals were then dissolved by pumping the tank contents in a recirculation mode back to the tank. While mixing, pH was adjusted to 11 using concentrated sodium hydroxide or sulfuric acid. The cleaning solution was then recirculated through the two pressure vessels at a flowrate of 4 gpm and a pressure of 30 psi. The concentrate and permeate flows were returned to the cleaning tank. After 60 minutes of pumping, the pump was stopped and the elements were soaked in the cleaning solution for another 60 minutes. To remove the residual surfactant solution from the system and membrane elements, the cleaning tank was emptied, inside tank walls rinsed several times with tap water, tank filled with tap water, and the tap water pumped through the vessels for 45 minutes. During the flushing step, concentrate and permeate was discharged to drain and new tap water added to the cleaning tank. The "compromised/fouled/cleaned element" test was conducted within 24 hours of membrane cleaning to minimize any biofouling.

A similar cleaning approach was employed at the NWR-WTP except that a commercially available cleaning solution was used as the cleaning agent (Ecolab high pH cleaner; 1:10 dilution with tap water). Figure 3.10 shows a picture of the cleaning skid used at the NWR-WTP.

Figure 3.10 Cleaning skid used in NWR-WTP

RESULTS AND DISCUSSIONS

The results and discussions are divided into two main sections based on the approach and objectives of the tests. The first section describes results for the SETU experiments in which the intact elements were tested for solute and surrogate removals using synthetic water. The second section describes results for the DETU, MSTU and MVTU experiments in which the effects of membrane or o-ring compromization, membrane fouling and cleaning, and the compromization location on surrogate removals were investigated using MFSE and CTRW polished by 10-μm cartridge filtration.

SETU Experiments

The results and discussions for SETU experiments are further divided into sections in terms of element and surrogate type.

ESPA1-Standardized Sodium Chloride Solutions

Solute rejection experiments using NaCl were conducted on seven ESPA1 4-inch diameter by 40-inch long spiral-wound elements. Results from each element are summarized in Table 3.6.

Table 3.6
Hydranautics ESPA1 Solute Results[a]

Element (Serial No.)	Feed Temperature (°C)	NaCl Feed Concentration (mg/L)	Feed Pressure (psi)	Flux (gfd)	Recovery (percent)	NaCl Rejection (percent)
A (A138081)	32.4	1740	94	21	23	98.7
B (A138082)	32.7	1620	97	21	24	98.8
C (A138084)	31.8	1720	95	21	23	98.6
D[b] (A138085)	31.8	1680	--	21	30	98.9
E (A138088)	32.3	1610	94	21	24	98.6
F (A138089)	32.2	1720	94	21	23	98.6
G (A138090)	30.9	1690	95	21	23	98.6

[a] Results shown were collected after 60 minutes of operation unless otherwise noted
[b] Results collected after 30 minutes of operation

Hydranautics' specified minimum rejection of NaCl is 99 percent based on a feed concentration of 1,500 mg/L NaCl at 150 psi feed pressure, 15-percent recovery, and 25 °C. Observed NaCl rejection was slightly less than minimum rejection for two reasons:

1) The SETU feed temperatures during testing were >25 °C. The rate of NaCl passage through RO and NF membranes increases approximately 3 percent per °C increase in feed temperature (salt passage equals 100-salt rejection, where salt rejection is expressed as a percent). For temperatures shown in Table 3.6, this equates to an 18-21% increase in NaCl passage relative to that at 25 °C, the temperature at which the performance at which the membrane rejection is rated. The temperature corrected (from observed to 25 °C) was estimated at 98.8 to 99.1 percent, higher than the minimum rated rejection.
2) Recovery was higher than 15 percent. At higher recovery, the osmotic pressure of the feed/concentrate is higher. At constant flux, this causes a reduction in NaCl rejection.

ESPA1-R-WT

Following completion of solute experiments, testing was conducted with R-WT testing using Element B (Serial No. A138082). Feed and permeate samples were collected every 15 minutes throughout the 120-minute test. Results from the test are shown in Figure 3.11. The experiment indicated virtually no decay of feed concentration throughout the entire 120-minute experiment. During bench scale testing at the University of Illinois, both microsphere and virus experiments repeatedly demonstrated substantial decreases in feed concentrations. Unlike microspheres and viruses (both particulates) R-WT is a dissolved, rather than a particulate substance, which are more capable of back diffusing into the bulk flow and not accumulating on surfaces. Feed pressure to maintain target flux and solute rejection remained constant at 98.8-98.9 percent throughout the 2-hour experiment indicating that the presence of low concentrations of R-WT in the membrane feed did not cause membrane fouling or change in solute rejection. Measurements of feed and permeate R-WT concentrations during this test showed a log rejection of approximately 2.7-3.0 logs.

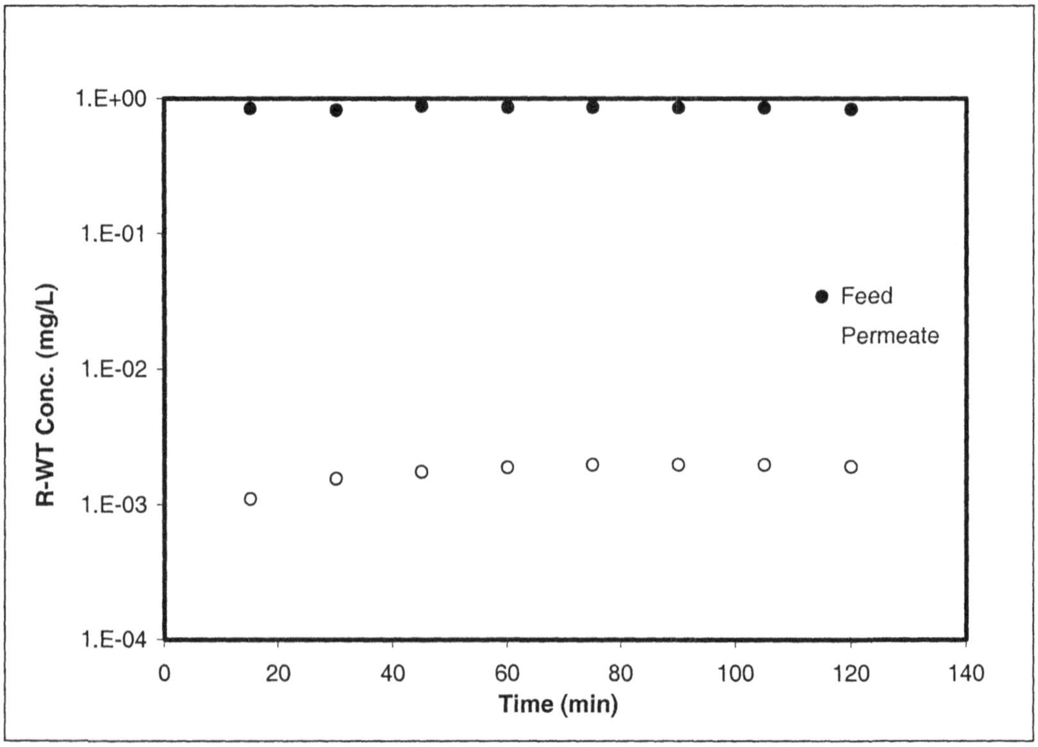

Figure 3.11 R-WT feed and permeate concentrations as a function of time for ESPA1 RO element

ESPA1-MS2 Phage

Element F (Serial No. A138089) was used to conduct the MS2 *phage* test. Based on bench results, it was determined that virus rejection could be affected (increased) by particulates present in the test loop accumulating on the membrane surface. Therefore a new membrane, previously tested for a short period on NaCl only, was used for this test.

Results from the *phage* experiment are shown in Figure 3.12 and Table 3.7. Initial seeding produced a concentration of 10^5 pfu/ml in the feed water. Permeate samples were analyzed and found to be absent of any phage indicating the ESPA1 element was achieving a >5- log rejection of *phage*. Re-seeding events at 65 and 145 minutes achieved feed concentrations of $3x10^5$ and 10^6, respectively. These data show complete removal of MS2 *phage* by the ESPA1 element at these feed concentrations (5-6 log virus removal).

Solute (NaCl) rejection increased throughout the virus experiments from 98.4 percent at t=20 minutes to 99.2 percent at t=155 minutes. This increase was caused by the higher flux at which the element was operated in the latter part of the test. Increased flux increases the rate of water flow through the membrane and dilutes the amount of salt passing through concurrently, which remains the same.

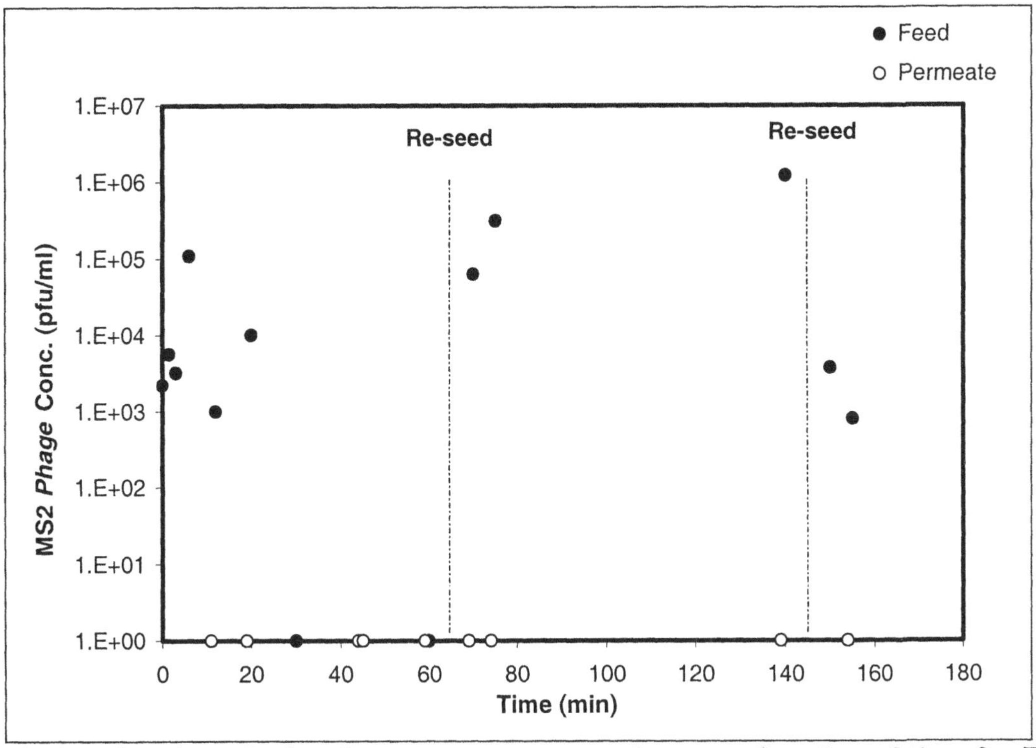

Figure 3.12 MS2 *phage* feed and permeate concentrations as a function of time for ESPA1 RO element

Table 3.7
ESPA1-MS2 *phage* experiment results

Time (min)	Feed *Phage* Concentration (pfu/mL)	Permeate *Phage* Concentration (pfu/mL)	NaCl Rejection (%)[a]	Permeate Flux (gfd)
0	8.5E+04	0	--	13
1.5	9.1E+04	0	--	13
3	8.3E+04	0	--	13
6	1.1E+05	0	--	13
9	1.1E+04	0	--	13
12	1.0E+03	0	--	13
20	1.0E+04	0	98.4	13
30	1.0E+00	0	98.5	13
45	1.0E+00	0	98.7	13

(continued)

Table 3.7 (continued)

Time (min)	Feed *Phage* Concentration (pfu/mL)	Permeate *Phage* Concentration (pfu/mL)	NaCl Rejection (%)[a]	Permeate Flux (gfd)
60	1.0E+00	0	98.7	15
70	6.3E+04	0	98.7	20
75	3.1E+05	0	98.7	20
140	1.2E+06	0	99.1	20
150	3.7E+03	0	99.2	20
155	8.0E+02	1	99.2	20

[a] NaCl feed concentration was measured to be 1520 mg/L at t=0.

ESPA1-Microsphere

Once virus experiments were completed with Element F, microsphere experiments were initiated using the same element. Feed concentration of microspheres during the ESPA1 experiment decreased from an initial value of 0.71 mg/L to 0.36 mg/L at 60 minutes (Figure 3.13). This "uptake" of microspheres either by the system components or the element was also demonstrated during bench-scale experiments.

Permeate sample analysis detected relatively high fluorescence that was initially attributed to the presence of microspheres. However, after further investigation, R-WT was also determined to be present in the permeate (by measurement using the field fluorometer). To better define this interference, calibration curves were developed for microsphere range of 1 to 100 ng/L with and without a background R-WT concentration of 30 ng/L. The results, shown in Figure 3.14, clearly illustrate the positive interference of R-WT on microsphere measurement. Concrete precautions are necessary to ensure that R-WT is to be used in a manner that will have minimum impact on microsphere measurements. As described previously, in subsequent testing (at both NWR-WTP and WF-21), R-WT was dosed *after* microsphere and MS2 *phage* during each experiment and membrane elements were flushed for 24 hours to remove residual R-WT prior to the conduct of a subsequent experiment. Flushing for this period was determined to be effective in reducing permeate R-WT to detection or below detection limits.

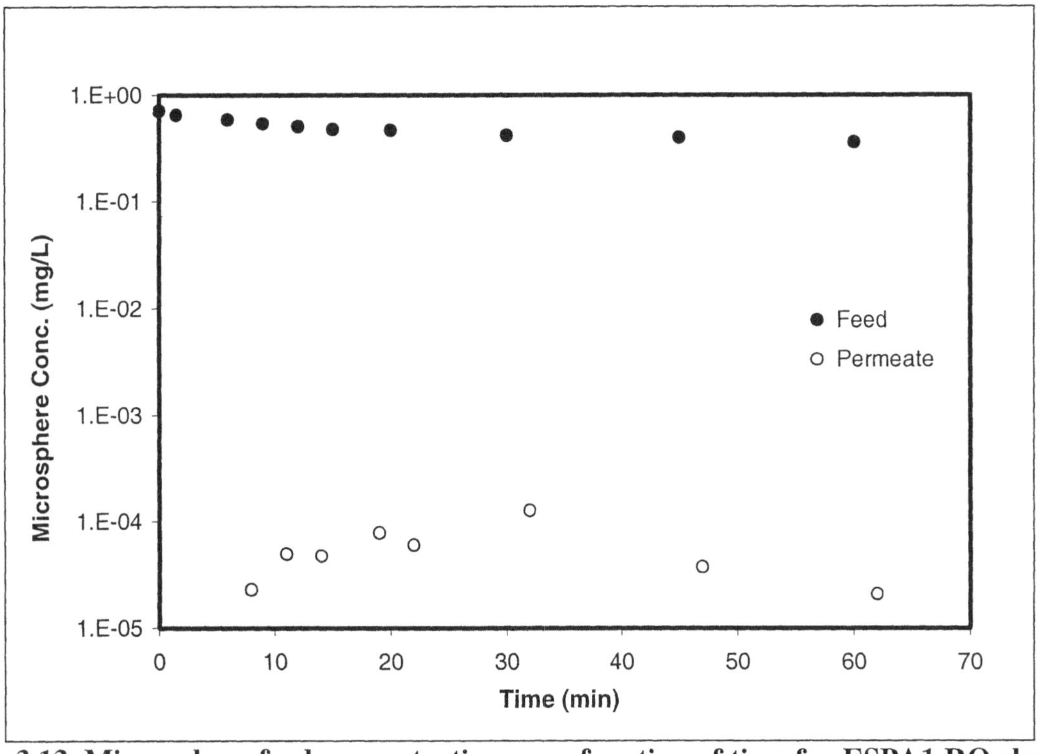

Figure 3.13 Microsphere feed concentrations as a function of time for ESPA1 RO element

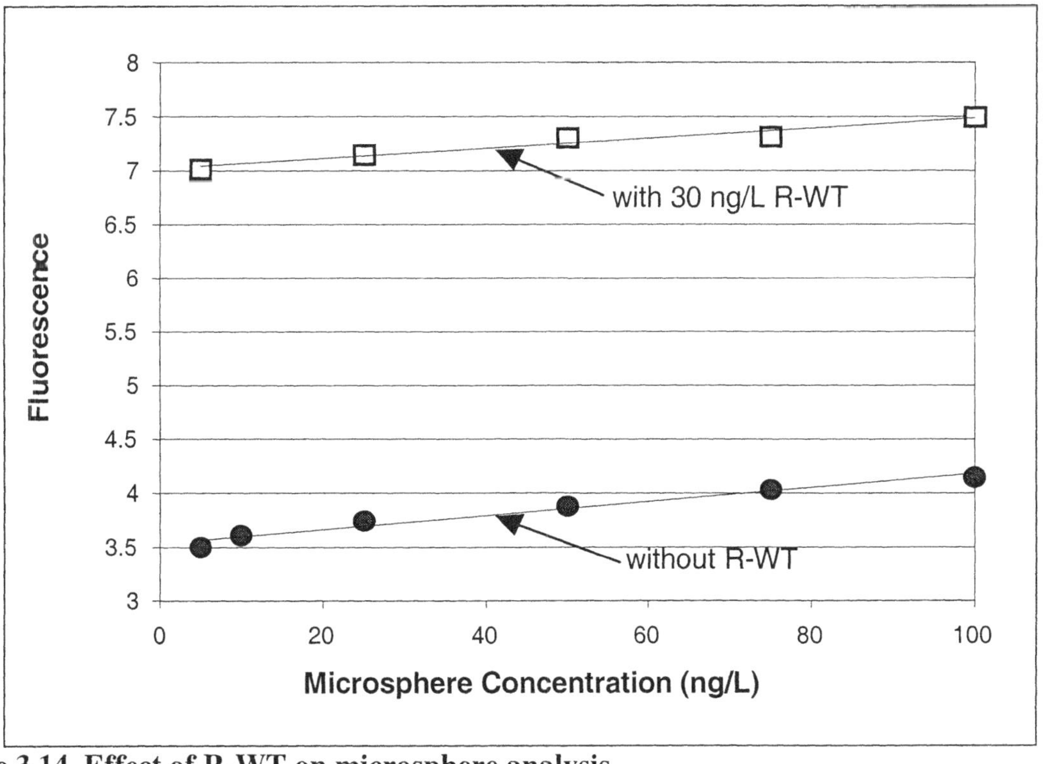

Figure 3.14 Effect of R-WT on microsphere analysis

Table 3.8
Fluid Systems TFCS Solute Results[a]

Element (Serial No.)	Feed Temperature (°C)	NaCl Feed Concentration (mg/L)	Feed Pressure (psi)	Flux (gfd)	Recovery (percent)	NaCl Rejection (percent)
H (582906)	30.0	575	68	20	19	96.1
I (583001)	30.4	575	70	20	19	95.9
J (583002)	29.9	567	70	20	19	95.2
K (583003)	30.2	581	70	20	19	95.9
L (583005)	30.5	581	71	20	19	96.5
M (583007)	29.5	563	72	20	19	95.6
N (583009)	29.4	571	70	20	19	95.6
P (583010)	30.3	579	73	20	19	97.1

[a] Results shown were collected after 60 minutes of operation unless otherwise noted

TFCS-Standardized Sodium Chloride Solutions

Solute testing was conducted on a total of eight TFCS 4-inch diameter by 40-inch length elements. Results from each element are summarized in Table 3.8.

Fluid Systems' specified design rejection of NaCl is 85 percent (measured as chloride ion) based on a feed concentration of 500 mg/L at 80 psi, 10-percent recovery, and 25 °C. Each of the membrane elements achieved a significantly greater rejection and were determined to be intact and void of any major imperfections. It was not determined why rejection was significantly greater than advertised, however it is not uncommon with NF membrane casting for higher NaCl rejection to occur. NF elements are manufactured to achieve a specific minimum hardness removal, as measured by rejection of $MgSO_4$, rather than NaCl.

TFCS-R-WT

The R-WT experiments were conducted with TFCS Element J (Serial No. 583002). Results from R-WT experiments are shown in Figure 3.15. R-WT feed concentrations declined substantially over the course of the 120-minute experiment. This was somewhat unexpected as the feed concentration of R-WT remained stable during ESPA1 element testing. By inference, the decline was attributed to uptake by the TFCS membrane as all other components in the SETU (and materials used in element construction) were identical between ESPA1 and TFCS tests. The R-WT uptake by the TFCS element may be due to differences of surface chemistry (charge, hydrophobicity, etc.) between the TFCS and ESPA1 active layer. Although the membranes in both elements comprise a polyamide active layer, there seems to be minor chemical compositional differences resulting in a differing degree of R-WT uptake.

Permeate R-WT concentration was about 7 μg/L resulting in log removal of 2.5 at the maximum R-WT feed concentration. This is less than the removal achieved by ESPA1. However, the log removal decreased as low as to 0.6 as the feed R-WT concentration declined during the experiment by nearly two orders of magnitude (Figure 3.15). As observed with the ESPA1 experiment, rejection of NaCl remained constant at around 96 percent throughout the

2-hour experiment indicating that R-WT did not impact inorganic solute rejection by the TFCS element.

TFCS-MS2 Phage

The MS2 *phage* experiment was performed using Element I (Serial No. 583001). Results are shown in Figure 3.16. A feed concentration of 4×10^5 pfu/ml of *phage* was achieved initially (t=0 min). The virus concentration decreased to 3×10^3 pfu/ml after 30 minutes of operation in the SETU. Permeate samples were analyzed and found to contain no viruses resulting in a >5.6-log rejection of *phage* based on initial feed level, values very close to those obtained by ESPA1 element. Reseeding events took place at 65 and 145 minutes resulting in feed concentrations of 6.8×10^4 and 2×10^3 pfu/ml, respectively. Permeate samples collected throughout the experiment contained no virus, except for one outlier at 150 minutes, which was most likely result of a sample contamination or analytical error.

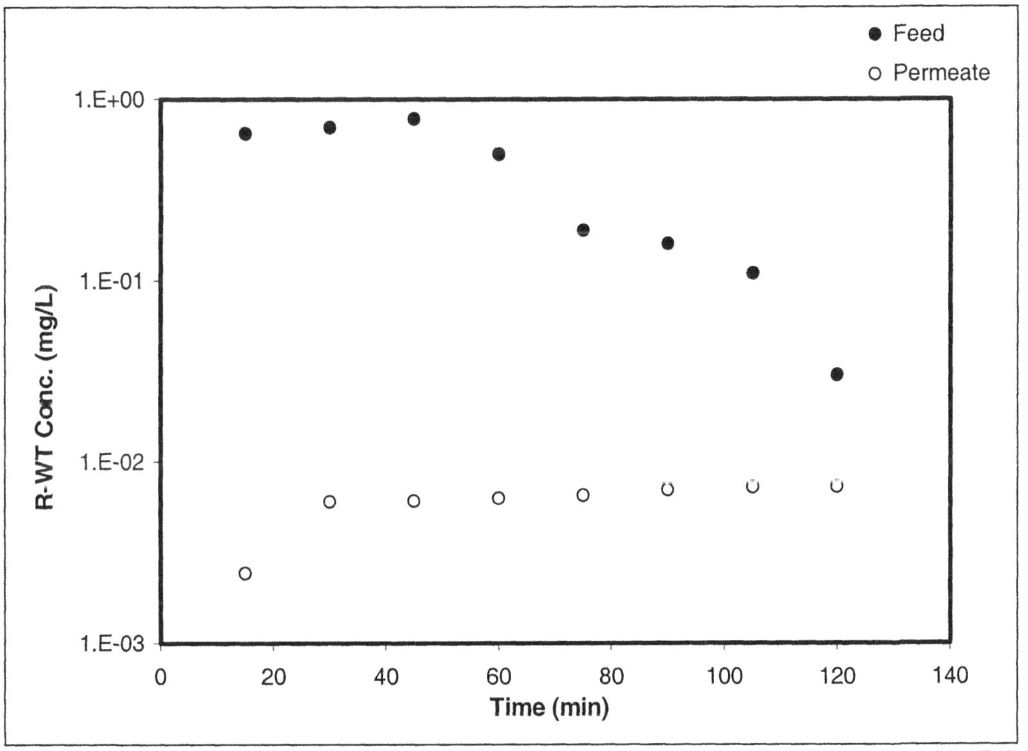

Figure 3.15 R-WT feed and permeate concentrations as a function of time for TFCS NF element

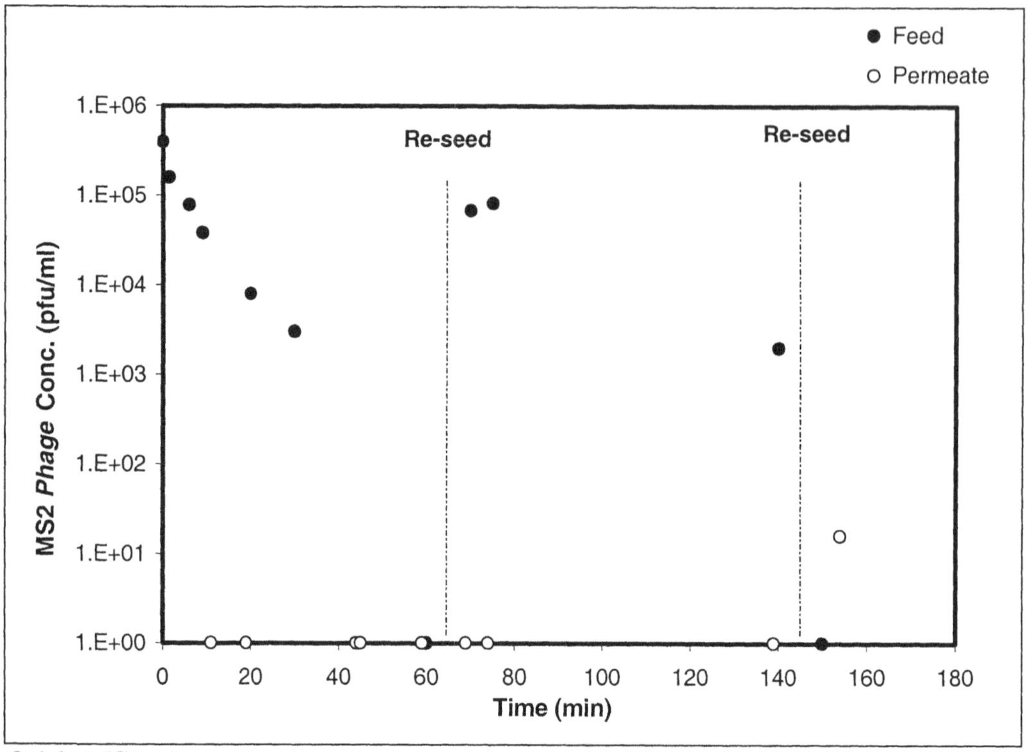

Figure 3.16 MS2 phage feed and permeate concentrations as a function of time for TFCS NF element.

TFCS-Microsphere

The microsphere experiment was conducted using Element J (Serial No. 583002), which was previously used for solute and R-WT rejection (Figure 3.17). The concentration of microspheres in the feed water decreased from 1.08 mg/L at t=0 minutes to 0.47 mg/L at t=60 minutes due to system uptake. The rate of system uptake corresponds with that observed during experiments with the ESPA1 membrane and indicates that microsphere uptake by the SETU is basically independent of any differences in surface chemistry between the ESPA1 and TFCS membranes. On the other hand, the results of the R-WT experiments showed the different extent of R-WT uptake by ESPA1 and TFCS elements. This suggests that surface adsorptive interactions of R-WT and microsphere with ESPA1 or TFCS elements may be different.

Similar to ESPA1 experiments, analysis of TFCS permeate samples detected the presence of microspheres. The positive detects may be the result of interference from residual R-WT, however the amount of interference could not be quantified. Based on the measured feed and permeate concentrations, TFCS removed between 3.5-4.2 logs of microspheres, similar to that obtained by ESPA1 element.

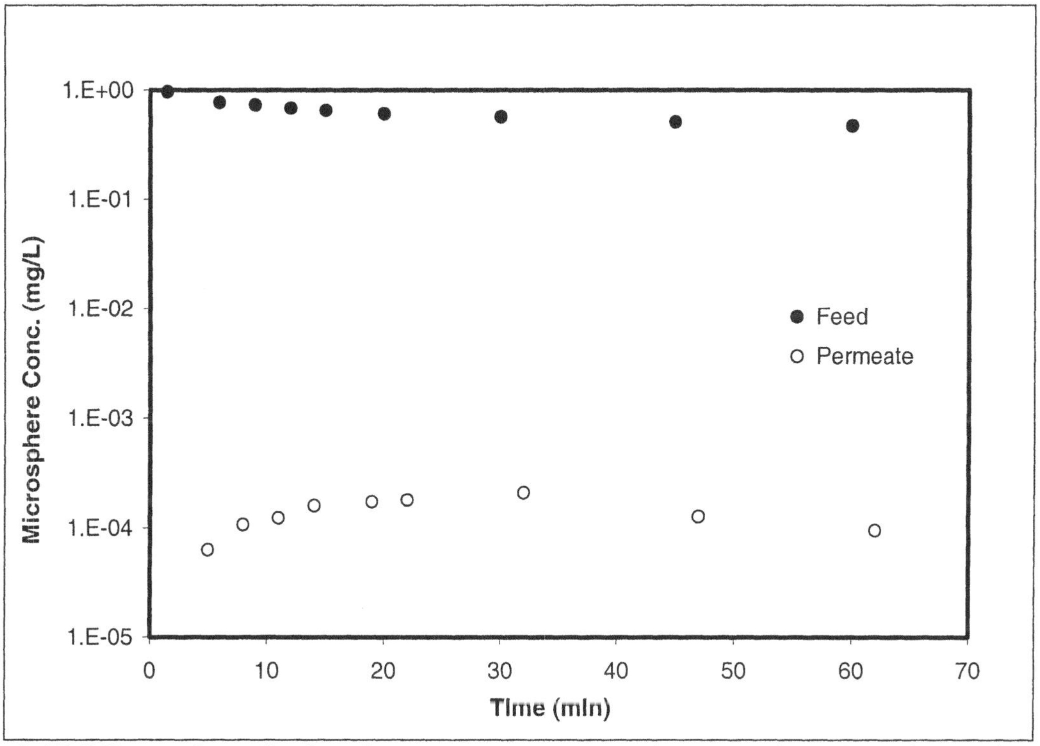

Figure 3.17 Microsphere feed and permeate concentrations as a function of time for TFCS NF element

DETU and MVTU Experiments

In DETU and MVTU experiments, the main objective was to investigate the impact of compromization (either of membrane and o-ring) and fouling and subsequent cleaning of membrane compromised elements on removal of microbial- and non-microbial surrogates. In addition, in MVTU experiments, the impact of the location of the compromised element and o-ring in a three-element vessel was also investigated. Before the discussion of results from DETU and MVTU, the following sections provide brief information about the membrane and o-ring compromization.

Membrane Compromization

Membrane compromization was achieved using a needle to induce a pinhole in the surface of one side of one leaf in the target element. This method of membrane compromization was chosen based on its successful application in the bench-scale experiments. The tape wrap on the outside of the element was first removed and the membrane leaves and associated feed spacer unrolled to access the selected leaf and membrane surface. The needle was then used to penetrate the membrane surface, with care being taken to penetrate fully through the complete membrane structure (polyamide active layer, polysulfone support layer and support fabric) without penetrating into the membrane structure on the facing side of the leaf. Following imperfection inducement, the leaves (and feed spacer) were re-rolled and secured with new tape wrap. Figure 3.18 shows the pinhole induced in one of the ESPA1 elements. The same pin and degree of force was applied for each membrane compromization in order to provide consistency. This is a hand technique, and as such, some variability in the size of the induced hole is to be expected.

O-ring Compromization

In the first compromised o-ring experiment, a crack was induced along the inner surface of the o-ring installed in the adaptor used to seal the element to the end cap at the lead end of the vessel. Testing showed that this degree of compromization was insufficient to cause a loss of integrity. Consequently, a greater degree of o-ring compromization was employed, namely the removal of small sections (i.e., 1, 2, or 4 mm) of the o-ring radially through the wall. Figure 3.19 shows the pictures of the intact and the compromised o-rings used in this study.

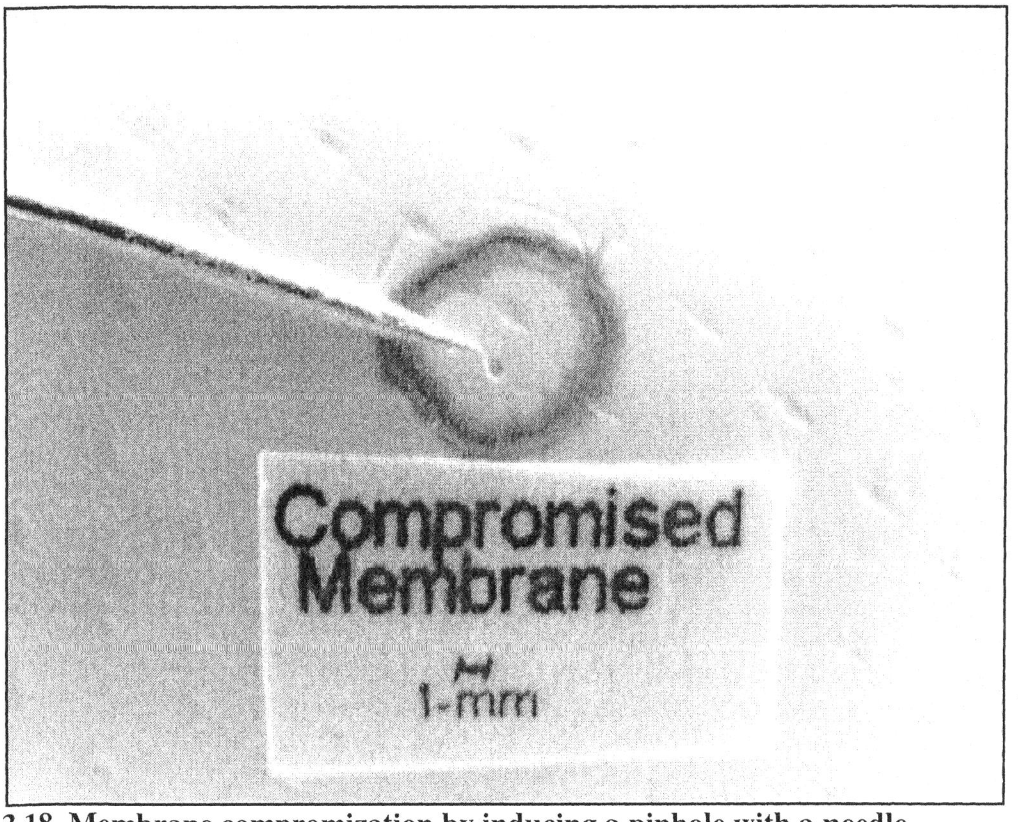

Figure 3.18 Membrane compromization by inducing a pinhole with a needle

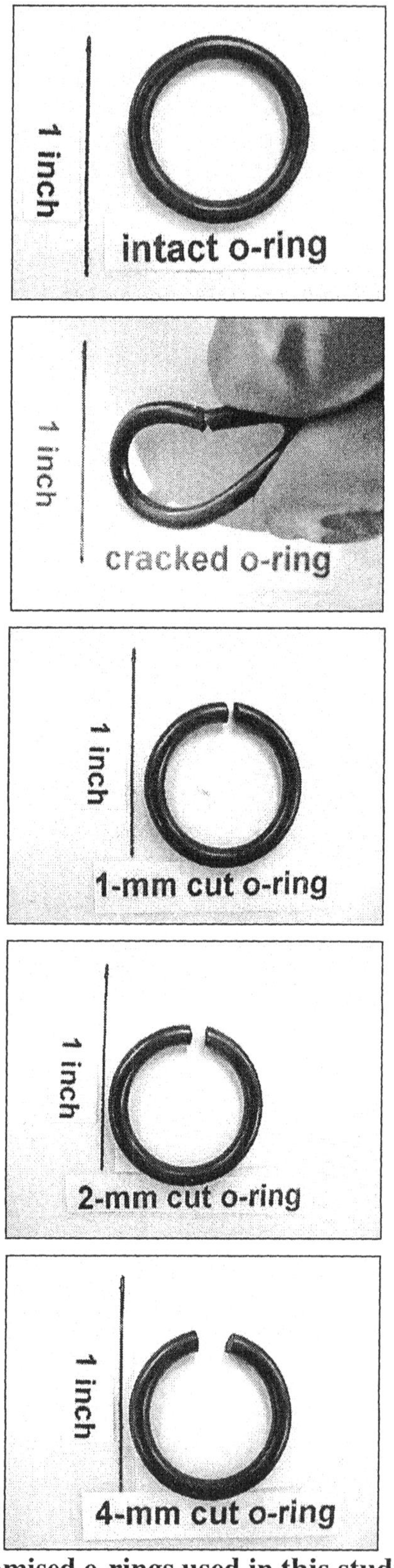

Figure 3.19 Intact and compromised o-rings used in this study

Table 3.9
Typical operating conditions for DETU and MVTU experiments

Test Unit	Feed Pressure (psi)	Recovery (%)	Permeate Flux (gfd)	Permeate Flowrate (gpm)
DETU	65-70	15	10.4	0.65
MVTU	60-65	45	12	2.2

DETU & MVTU Results

Results from DETU and MVTU experiments are discussed in subsections based on element type and surrogate tested. DETU experiments were conducted at the WF-21 using MFSE, while MVTU experiments were conducted at the NWR-WTP using CTRW.

ESPA1-R-WT Testing. Of the three surrogates employed in the pilot testing, R-WT was studied most extensively because of the ability to measure impacts real time using the field fluorometer. Consequently, the R-WT results are presented first.

R-WT experiments were conducted with ESPA1 elements having intact or compromised membrane or compromised o-ring. Figures 3.20 and 3.21 show the log R-WT removals obtained with ESPA1 elements at different compromization and at different degrees of membrane fouling for DETU and MVTU, respectively. Operating conditions for the two sets of experiments are shown in Table 3.9.

Intact membranes and o-rings. With the DETU experiments, 4-log R-WT removal was achieved based on a target feed concentration of 1 mg/L (Figure 3.20). With the MVTU, R-WT removal was more variable, ranging from 3.5 to 5.3 logs (Figure 3.21). For DETU, the removals were relatively constant as a function of time. In contrast, removals declined over the 45-min sampling period for the MVTU. The temporal decline during MVTU tests was consistently observed when initial removals were high (>4 logs). It is hypothesized that the decline is caused by adsorption of the dye molecule into the membrane structure and subsequent "bleeding" of the dye through the membrane, causing a lower of rejection. It is not clear, however, why the adsorption would occur to a greater degree with CTRW or why it would be more pronounced in a 3-element array versus a single element. The latter difference could be related to the effect of recovery (higher concentration of dye in the tail end of the 3-element vessel). Notwithstanding this difference, these results show that similar log removals of R-WT were demonstrated by the ESPA1 RO membrane on the two different feedwater matrices.

Membrane compromization. Compromising the ESPA1 element with a pinhole significantly decreased log removal, from 4 to 2.2 logs in both DETU and MVTU tests (Figures 3.20 and 3.21). This result suggests pinhole size was similar for both tests (estimated at 300-400 μm, based on the digital picture shown in Figure 3.18). Significant reductions in log removals were also found with flat-sheet membrane coupons using the same pinprick technique in bench-scale experiments. Interestingly, in the MVTU experiments, log removal increased significantly when the compromised ESPA1 element was switched from lead position (2.2 log removal) to trailing position (4.3 logs initially, declining to 3.8 logs after 45 minutes). In the trailing position, the element would have been subjected to a higher concentration of R-WT (approximately 1.5 times feed concentration versus 1.1 times for lead position), but a lower pressure differential across the pinhole. Apparently, lower pressure differential resulted in less passage of the dye through the pinhole providing larger log removals. Overall, these results indicate the importance of maintaining membrane integrity in full-scale applications in order to achieve high level of viral removals as dictated by the stringent regulations.

Fouling and cleaning of compromised membrane. For both DETU and MVTU experiments, fouling the compromised ESPA1 element increased the log removals significantly, even somewhat higher than those obtained with the intact element (Figures 3.20 and 3.21). This finding suggests that fouling the elements may effectively clog the imperfections on active layers (i.e., the pinhole in this study) possibly by forming a cake layer on them. As evident with removals larger than those of intact elements, these cake layers may cover the inherent imperfections on the elements and may even increase the treatment efficiency by producing an additional adsorptive barrier. On the other hand, chemically cleaning the compromised and fouled ESPA1 elements decreased the log removals, possibly because of the sheared foulant cake layer during cleaning. However, the extent of reductions in log removals was different between DETU and MVTU. Cleaning the element in DETU decreased the log removals back to levels almost the same found for compromised element showing the complete removal of the foulant layer. For MVTU however, the cleaning was less effective. There may be two reasons for this. One is that the commercial cleaning mixture used for MVTU test was not as effective. Second, the characteristics of the formed cake layers may be quite different due to the different nature of the feedwaters, i.e., the cake formed in MVTU tests could be relatively irreversible.

O-ring compromization. Cracking the o-ring did not increase the passage of R-WT in DETU experiments (Figure 3.20), a result consistent with the bench-scale findings with MS2 *phage* and microspheres. Apparently, the force generated by the pressure differential across the o-ring (between feed and permeate) served to compress the o-ring at the crack and provide an effective "dynamic" seal against dye passage. On the other hand, removing small sections from the o-ring did result in decreased log removals in both DETU and MVTU tests (Figures 3.20 and 3.21). As expected, the larger the removed section, the greater the reduction in log removals was: 1-, 2- and 4-mm removed sections decreased log removals to 3.2, 1.3 and 0.6, respectively. For the 4-mm removed section o-ring experiment in DETU, at the permeate flow rate of 0.65 gpm, 0.6 log removal (<90% removal) would mean that more than 10% of the permeate flow (~0.07 gpm) would be derived from the passage of feedwater across the o-ring defect. Based on a target of 50% reduction in log removal by o-ring compromization, the 2-mm removed section was selected for subsequent compromization experiments conducted with the MSTU and MVTU.

While a 2-mm cut o-ring in the lead adaptor position significantly decreased the log R-WT removals, the removals were comparable to those of the intact o-ring experiment when the compromised o-ring was positioned in the trailing adaptor (Figure 3.21). Although it is difficult to explain this observation, one plausible explanation would be that the compromised o-ring was placed in the trailing adaptor in such a way that it provided an effective barrier by closing the gap in the removed section of the o-ring. The sensitivity of the removals to the degree of o-ring compromization is also evident from the different log removals found for DETU and MVTU tests at the same condition of 2-mm cut: 1.3 and 2.8 log removals in DETU and MVTU, respectively.

Other observations. For both DETU and MVTU tests, slight reductions in R-WT log removals were observed with increasing run time (up to 45 minutes). This is probably due to the adsorption of R-WT molecules within the polyamide rejecting layer, and its subsequent diffusion into the permeate. No attempt was made to do a fit of the data to determine if this reduction would continue or diminish with time, however, such behavior suggests that use of the R-WT as a viral surrogate may be done using short-term seeding.

Rejection of conductivity (Tables 3.11 and 3.13 for DETU and MVTU, respectively) and TDS (Table 3.12 for DETU) during 45 minutes of R-WT injection was constant for all ESPA1 experiments and did not differ from that observed prior to R-WT dosing. Similar results were also found for MS2 *phage* and microspheres, indicating that the presence of low concentrations of the surrogates did not reduce inorganic ion rejection either through cake layer formation (*phage* or microspheres) or adsorption within the membrane (R-WT). The rejection data of conductivity and TDS in Tables 3.11-3.13 will be discussed in detail in a separate subsection.

The feed R-WT concentration was very stable during the 45 min of testing with a percent relative standard deviation (coefficient of variation, %CV) of less than 3% between the measured feed concentrations. This suggested that the dosing pump delivered constant flow during the testing and that only minimal amounts of R-WT were "lost" through sorption on process equipment surfaces.

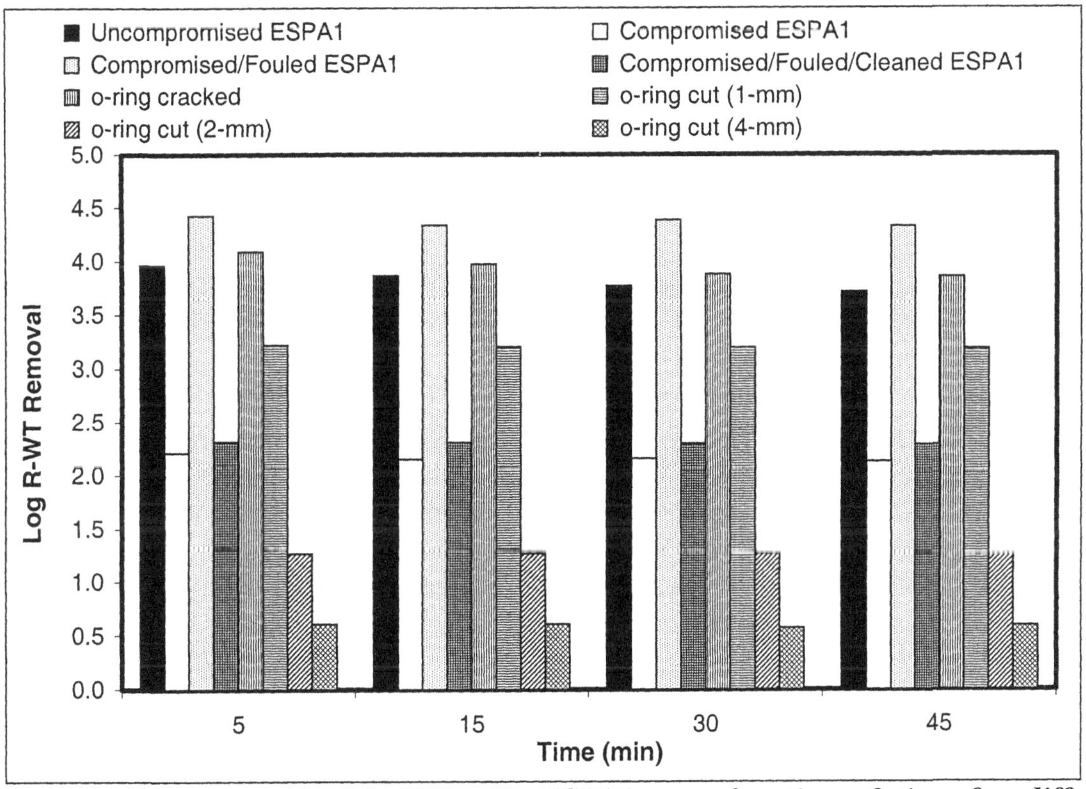

Figure 3.20 Log removal of R-WT with ESPA1 as a function of time for different experimental conditions in DETU

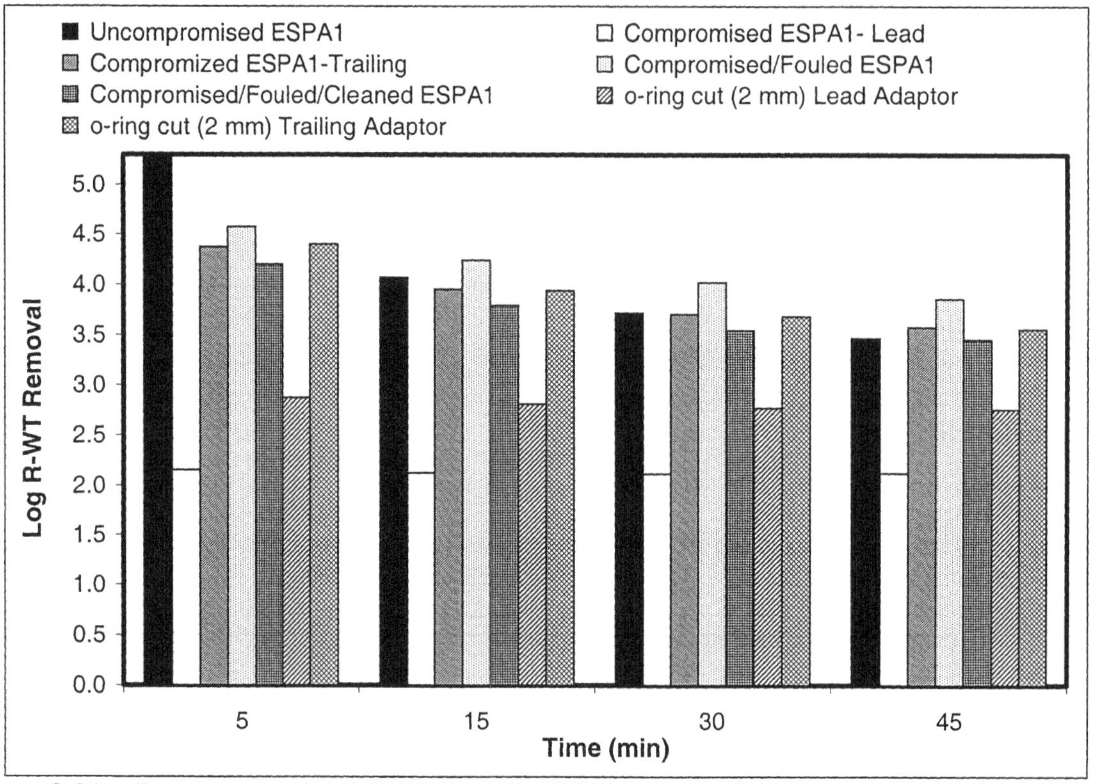

Figure 3.21 Log removal of R-WT with ESPA1 as a function of time for different experimental conditions in MVTU

ESPA1-MS2 Phage Testing. As a biological surrogate for enteric viruses, MS2 *phage* was tested for its removal by ESPA1 elements having intact or compromised membrane or element/vessel system having a single compromised o-ring. Figures 3.22 and 3.23 summarize the log removals obtained in DETU and MVTU, respectively.

Intact membranes and o-rings. Complete removal was achieved (log 6.8) with the intact ESPA1 in DETU, similar to that found in bench-scale testing and in SETU experiments. The log removal however was only 5.3 in MVTU experiment. The feed MS2 concentrations were between 10^6-10^7 pfu/ml in all the experiments, indicating that low levels of virus (about 2.4 logs) was passing the intact ESPA membrane during the MVTU experiment.

Membrane compromization. Compromising the ESPA1 element with a pinhole also significantly decreased the log MS2 removals from 6.8 to 2.9 in DETU, and from 5.3 to 2.3 (lead position) and from 5.3 to 4.2 (trailing position) in MVTU. Similar degrees of reduction were observed for R-WT. Consistent with R-WT results, the impact of membrane compromization on *phage* removal was less when the compromised element was in the trailing position.

Fouling and cleaning of compromised membrane. For both DETU and MVTU experiments, fouling the compromised ESPA1 element increased the MS2 log removals significantly, consistent with the R-WT results (Figures 3.22 and 3.23). For MVTU, the fouling even increased the removals more than those obtained by the intact membrane. These findings suggest that the fouling layer is quite effective in clogging both the larger induced imperfection and smaller imperfections intrinsic to membrane casting that were most likely present in the element used in the MVTU intact test. Chemically cleaning the compromised and fouled ESPA1 elements decreased the log removals, as found with R-WT. However, again, the extent of reductions in log removals was different between DETU and MVTU; DETU cleaning being more effective reducing the removals back to those of compromised element.

O-ring compromization. As found for R-WT, the cracked o-ring did not result in MS2 passage, consistent with the finding of bench-scale experiments (Figure 3.22). Interestingly, removing a 1-mm chunk from the o-ring also did not change the log MS2 removals. The variation in removals among the uncompromised membrane and compromised o-ring tests are most likely caused by experimental and analytical variations between the experiments. This result differs from that for R-WT, and suggests that the smaller dye molecules may be better able to penetrate the partially compromised o-ring. One would anticipate that log removals would decrease with removal of larger sections of o-ring, although this was not tested for MS2 *phage* due to time constraints.

For MVTU tests, as with R-WT, the 2-mm cut o-ring in the lead adaptor position significantly decreased the log MS2 removals, while little in any loss of removal was observed when in the trailing position. Again, one plausible explanation would be that the compromised o-ring was placed in the trailing adaptor in such a way that it provided an effective barrier.

Other observations. The slight reductions in the log removals observed for R-WT as a function of time were not found for MS2 both in DETU and MVTU tests. The log removals were nearly constant during the 45 min of injection. As MS2 is a discrete particle not capable of dissolving into the RO rejecting layer, a reduction in removal would not be expected. As observed for R-WT, the feed MS2 concentrations to test units were very stable during dosing, with a percent relative standard deviation of maximum 19%, although the majority of the runs had values around 5-11%. The larger percent relative standard deviations (compared with R-WT) may be due to the fact that the feed MS2 samples require significant dilution prior to enumeration, whereas the R-WT feed concentration is measured directly. Dilution increases error. Nevertheless, the relatively constant feed MS2 concentrations over injection time suggests that MS2 *phage* was not lost in the DETU and MVTU through mechanisms such as sorption. Feed MS2 concentrations were very close to the target level of 10^7 pfu/ml, suggesting that the use of sodium bisulfite to quench chlorine in the MFSE was effective in preventing *phage* inactivation.

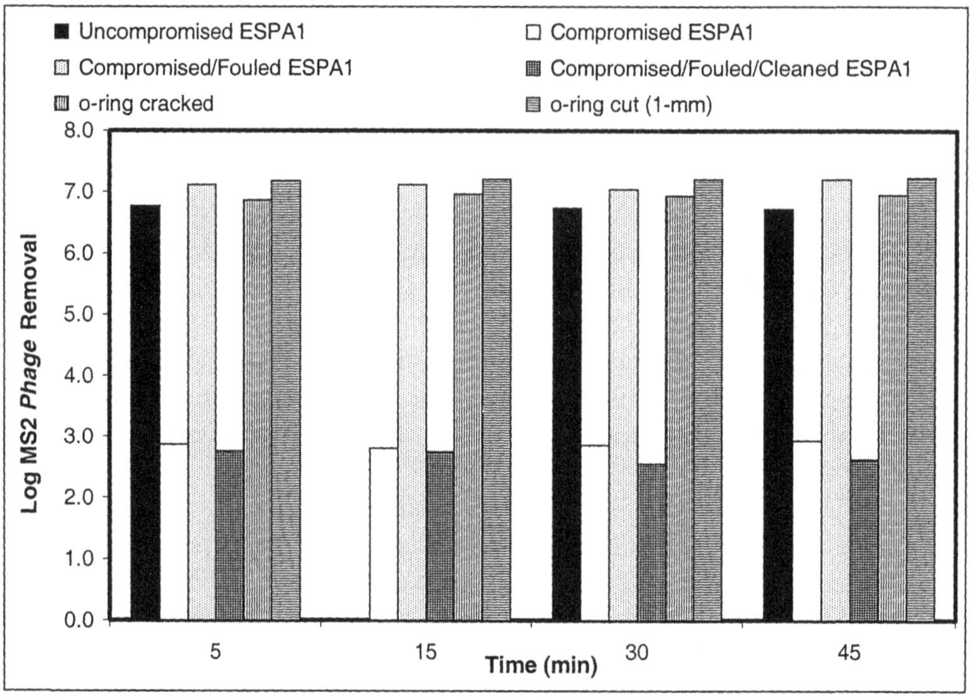

Figure 3.22 Log removal of MS2 *phage* with ESPA1 as a function of time for different experimental conditions in DETU (Log removal for uncompromised element could not be calculated at t=15 minutes because permeate sample was broken during the shipment; feed sample at this sample time was similar to that for other times).

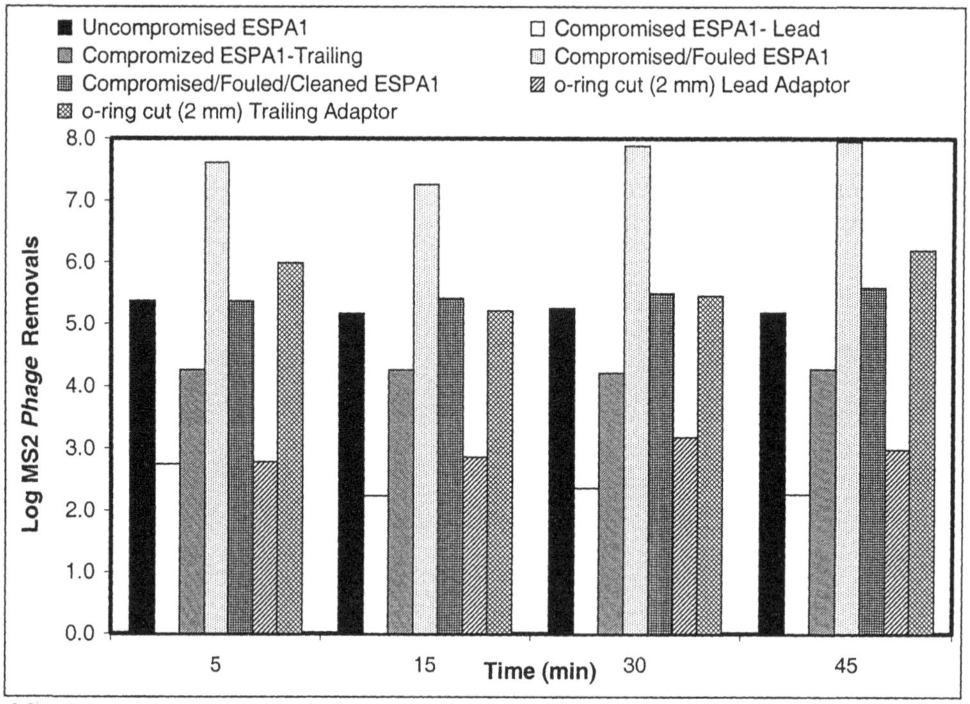

Figure 3.23 Log removal of MS2 *phage* with ESPA1 as a function of time for different experimental conditions in MVTU

ESPA1-Microsphere Testing. The log microsphere removals obtained at different experimental conditions are shown in Figures 3.24 and 3.25 for DETU and MVTU, respectively.

Intact membranes and o-rings. Log removals of about 4 were measured for the intact ESPA1 element in DETU and MVTU tests, the same value found for R-WT with both having a feed concentration of approximately 1 mg/L. This suggests that the removal mechanisms of R-WT and microspheres by ESPA1 may be similar (i.e., size exclusion) even though spheres are in particulate form and R-WT is a dissolved substance. In the bench-scale experiments, the microsphere log removals were found to be 5-6. However, the feed concentration employed was 10 mg/L, a dose that was prohibitively expensive for pilot scale work. As such, one would expect that the uncompromised ESPA1 would demonstrate >4 log microsphere removal at higher feed concentrations.

Membrane compromization. Membrane compromization decreased sphere log removals from 4.0 to about 2.2-2.3 in both DETU and lead-element compromised MVTU experiments (larger removals were observed for compromised trailing element). These results are in good agreement with the results for R-WT. In addition, as discussed in the previous section, MS2 removals decreased from 5.3-6.8 to 2.3-2.9 with membrane compromization. Similar degree of reductions in surrogate removals may suggest that the pinholes induced on the ESPA1 elements allowed passage relatively independent of surrogate type and physicochemical characteristics. This further suggests that the passages of enteric viruses through the membrane imperfections (i.e., a pinhole) may be predicted by biological and non-biological surrogates as used in this study, as long as the size of a pinhole is large enough and does not cause a steric hindrance to passage. This is an important finding, in that a non-particulate surrogate (R-WT), with its attendant advantages of low cost and ease of use, could be used as a particulate (viral) surrogate.

Fouling and cleaning of compromised membrane. For both DETU and MVTU experiments, fouling the compromised ESPA1 element increased the microsphere log removals significantly (from about 2.3 log to about 4.2 logs), consistent with the R-WT and MS2 *phage* results (Figures 3.24 and 3.25). Chemically cleaning the compromised and fouled ESPA1 elements decreased the log removals. As with thc R-WT and MS2 finding, chemical cleaning was more effective in removing the foulants present in the MFSE that increased log removal than for foulants present in the CTRW.

O-ring compromization. The log removal data for the o-ring crack experiment was not as conclusive as those obtained from R-WT and MS2 experiments (Figure 3.24). A high level of variability in log removal was observed (ranging from 2.6 and 4.3). This may be due to analytical variations associated with the low-level measurement of permeate samples, which had concentrations close to the detection limit of the fluorometer (about 50 ng/L). Three of the permeate samples had concentrations less than 500 ng/L, providing log removals around 3.5, relatively close to those found for the intact o-ring case. However, one permeate sample had a concentration of 1,940 ng/L, which resulted in a much lower log removal of 2.6. This type of variation was not observed with the other surrogates. It may be also possible that the high concentration permeate sample was contaminated during collection or was impacted by the presence of residual R-WT from the prior experiment. The latter is not considered plausible based on the consistent reduction in R-WT levels in the RO and NF permeates to non-detect levels (i.e., less than 20 ng/L) after the 24-hour flushing period.

For MVTU tests, the 2-mm cut o-ring in the lead adaptor position significantly decreased the log microsphere removal, consistent with the R-WT result. As was observed for R-WT also, locating the cut o-ring in the trailing position had much less impact on log removal reduction (from 4.4 to 3.8 logs) than in lead position (to 2.7 log).

Other observations. The log removals were nearly constant during the microsphere injection, which was expected as spheres, like MS2 *phage*, were discrete particles not capable of dissolving into the membrane rejecting layer. Feed microsphere concentrations were also stable during the injection with a percent relative standard deviation of less than 9% in DETU and MVTU experiments, demonstrating the consistency of dosing solution strength and dosing flowrate.

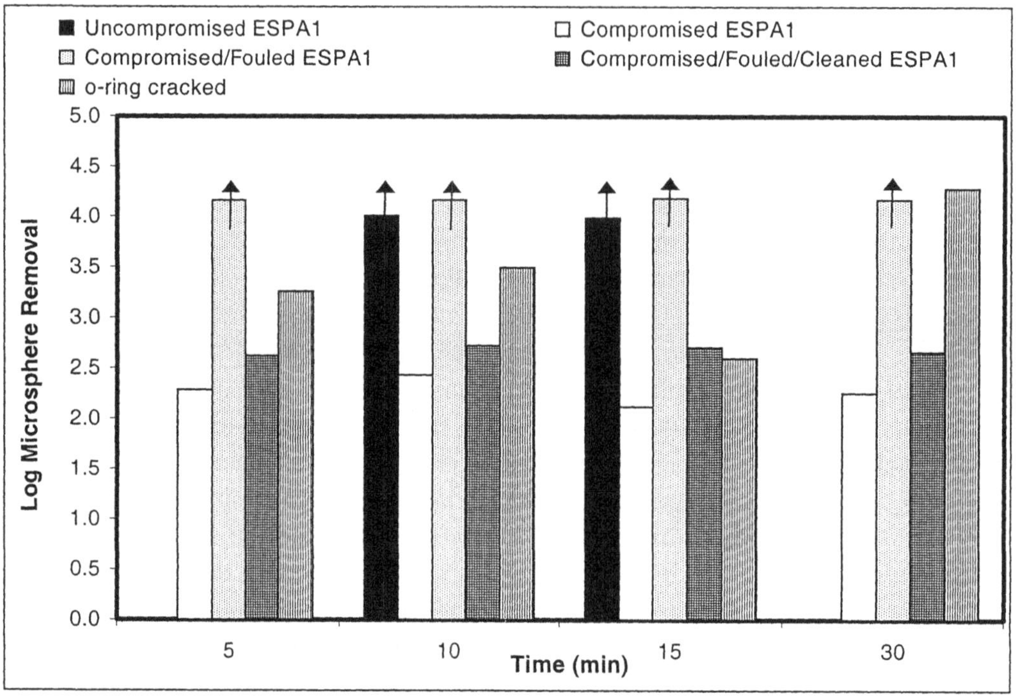

Figure 3.24 Log removal of microsphere with ESPA1 as a function of time for different experimental conditions in DETU (For the uncompromised ESPA1, feed sample at t=30 min and permeate sample at t=5 min were broken during shipment; thus the removals could not be calculated at these sampling times. However, the concentration of the permeate sample at t=30 min was about at detection limit very close to those of permeate samples at other sampling times. Similarly, feed concentrations were also very consistent as a function of dosing time. Therefore, the log removals of the uncompromised ESPA1 at t=5 and 30 min would be expected to be similar to those measured at t=10 and 15 min. Log removals of >4 (indicated by the upward-oriented arrow above the bar) are shown for the uncompromised and fouled element since the permeate concentrations were below quantification limit of 50 ng/L).

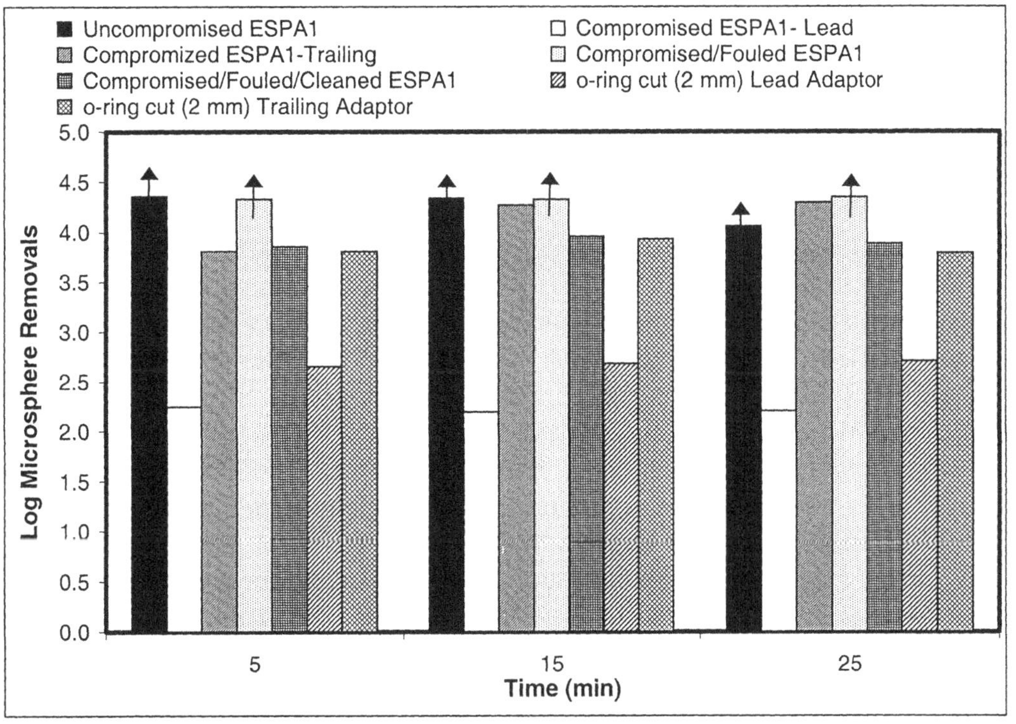

Figure 3.25 Log removal of microsphere with ESPA1 as a function of time for different experimental conditions in MVTU (Log removals of >4-4.2 (indicated by the upward-oriented arrow above the bar) are shown for the uncompromised and fouled element since the permeate concentrations were below quantification limit of 50 ng/L).

TFCS-R-WT Testing. *Intact membranes and o-rings.* In the first experiment of DETU (Experiment W1, Table 3.2) using an intact TFCS element, lower-than-expected log removals were found. In fact, greater log removal of R-WT was measured for the compromised TFCS element (membrane pinprick, Experiment W2) than for the uncompromised element tested in W1. It should be noted that the element used in W2 was a different element than used in W1. These findings suggested that there may be intrinsic imperfections in the TFCS membrane (or element) resulting from the manufacturing process. These imperfections are not distinguishable using normal manufacturer's QA/QC "wet testing" which simply measures percent rejection of magnesium sulfate at <99% (no differences were observed in conductivity rejection between Tests W1 and W2). To test this hypothesis, two additional uncompromised TFCS elements (as received from the manufacturer) were challenged with R-WT and MS2 *phage*. The new tested elements exhibited similar R-WT removals of around 3 logs, compared with 2.5 logs for the first element tested, confirming that the first TFCS element and/or membrane had already some degree of compromization inherent with its manufacturing process (Figure 3.26). The difference was much more dramatic for MS2 *phage* as will be discussed in the following section. As a result, the compromization test (W2) was repeated by compromising one of the intact elements (TFCS#3 in Figure 3.26). The removal performances of intact ESPA1 and TFCS elements are specifically compared and discussed in a following separate section.

Figures 3.27 and 3.28 show the removal of R-WT at different experimental conditions in DETU and MVTU, respectively. For DETU, removal ranged from 3 logs initially to 2.75 logs at 45 minutes. For MVTU, removals ranged from 4 log initially to 2.9 logs at 45 minutes.

Membrane compromization. For DETU, TFCS membrane compromization also reduced removal, from 3.0 to 1.5 logs, or nearly 50% (Figure 3.27), which is consistent with the degree of reduction observed for all the surrogates tested with the ESPA1 element. This result is consistent with an imperfection the size of that caused by the pinhole (approximately 300-500 μm in diameter), where advective flow (roughly equivalent for both membranes at equal flux rates) through the hole would result in equivalent passage of both virus-sized particles and dye-size molecules. For MVTU, the degree of TFCS membrane compromization was not sufficient resulting in small reductions in log R-WT removals compared with those of intact TFCS (Figure 3.28). Since membrane compromization by inducing a pinhole is a hand-technique, such variations between the performance of compromised membranes would be expected. For the MVTU, no significant difference between the performance of the lead and trailing positioned compromised TFCSs elements were found, however, given the limited impact of compromization on removal, this is not unexpected.

Fouling and cleaning of compromised membrane. For both DETU and MVTU experiments, although fouling the compromised TFCS element increased the R-WT removals, the extent of increase was not as large as those found for ESPA1 element (Figures 3.27-3.28). Especially for DETU, where compromization had the greatest impact, did not help to increase the log removals to levels observed for the intact element. The lack of impact is attributed to a lesser amount of fouling on the TFCS element. Digital pictures taken of the surface of both elements after fouling show a significant foulant layer formed on the ESPA1 membrane but very little on the TFCS membrane (Figures 3.29-3.30). Chemically cleaning the compromised and fouled TFCS elements decreased the log removals back to levels of compromised TFCS. Unlike the ESPA1 MVTU experiment, where cleaning did not reverse foulant impacts, it appears that the thinner foulant formed on the TFCS membrane was more readily removed by chemical cleaning.

O-ring compromization. Cracking the o-ring caused a small decrease in R-WT removal, from 2.9 to 2.6 log (Figure 3.27), a result not observed with the ESPA1 element. The difference could reflect the lower pressure differential applied to the o-ring (TFCS element operated at lower feed pressure than ESPA1), which may not have caused the two halves of the o-rings to seal as effectively at the crack. However, this hypothesis cannot be confirmed and other factors may explain the difference, including slight differences in the seal between the o-ring and the permeate tube, based on differences in permeate tube design. Consistent with the ESPA1 results, however, was the impact of removing o-ring sections both for DETU and MVTU (Figures 3.27-3.28); log removal decreased with increasing size of the section removed. In terms of the impact of compromised o-ring location in MVTU tests, removals were slightly greater for lead versus trailing position. This result is in contrast to those for the ESPA1, where trailing positioned 2-mm cut o-ring consistently exhibited larger removals for all the three tested surrogates.

Other observations. As observed for the ESPA1 element, the log removals of R-WT decreased by time, indicating dye adsorption within the polyamide rejecting layer and subsequent release into the permeate. However, the reductions in removals by time were more significant with TFCS than ESPA1 (almost for all experimental conditions independent of compromization types, etc.), further supporting the previous finding with 24-hour flushing test that TFCS membrane adsorbs more R-WT most probably due to differences in surface chemistry. On the other hand, the lower log removals for the TFCS NF element compared with the ESPA1 RO element (4 logs) suggests that the lower molecular weight cutoff (MWCO) of the RO membrane provides greater R-WT removal. This characteristic distinguishes a dissolved surrogate, like R-WT, from a particulate surrogate, like microspheres, where the difference in MWCO of the two membranes would not influence removal.

Rejection of conductivity (Tables 3.11 and 3.13) and TDS (Table 3.12) during R-WT injection was constant for all TFCS experiments and did not differ from that observed prior to R-WT dosing. Similar results were also found for MS2 *phage* and microspheres, indicating that the presence of low concentrations of the surrogates did not cause impact on membrane inorganic ion rejection.

In terms of the variations in the feed concentrations during dosing, the same relative standard deviations apply as found for ESPA1 experiments since both elements were tested concurrently and received the same feedwater.

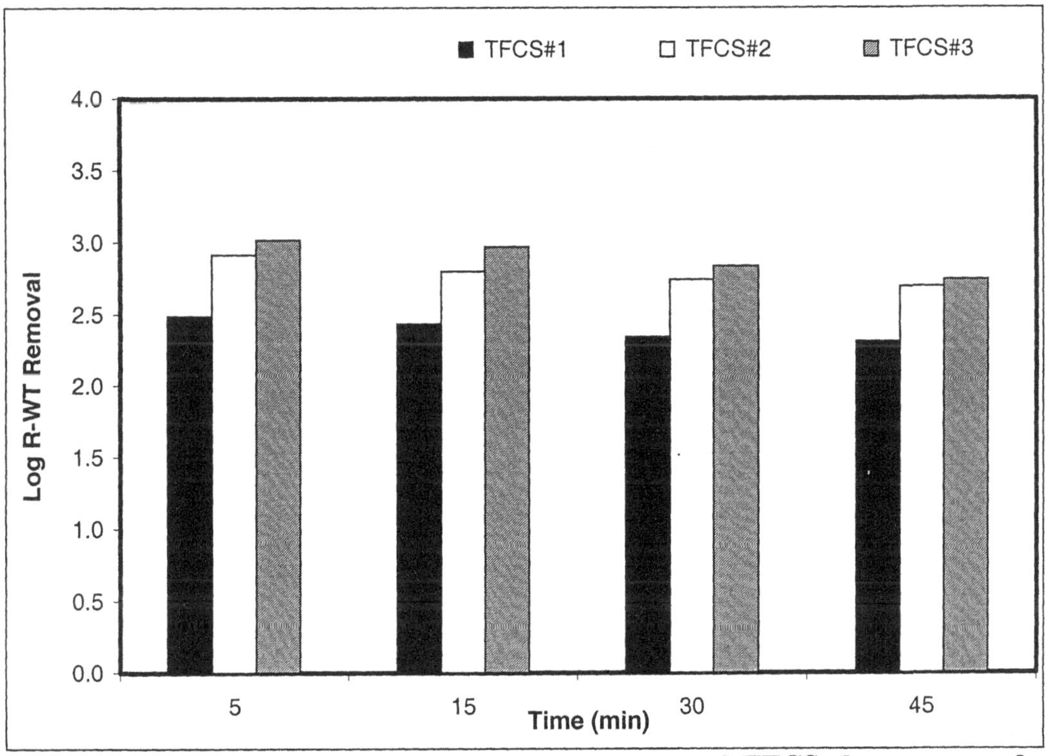

Figure 3.26 Log removal of R-WT for three uncompromised TFCS elements as a function of time in DETU

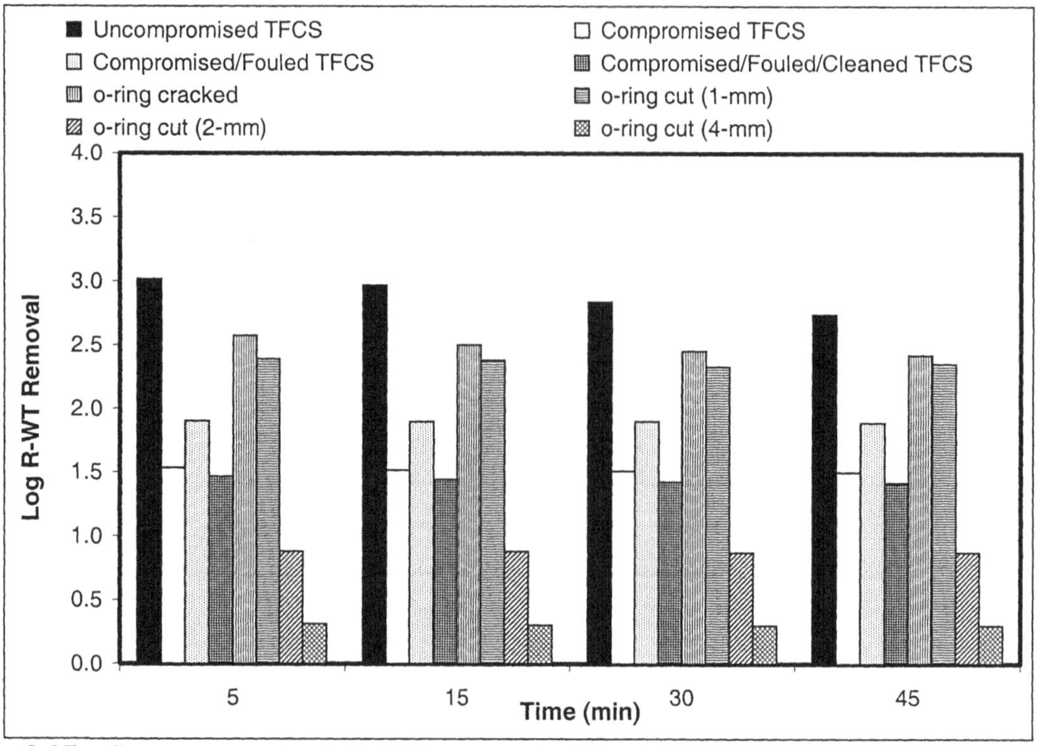

Figure 3.27 Log removal of R-WT with TFCS as a function of time for different experimental conditions in DETU

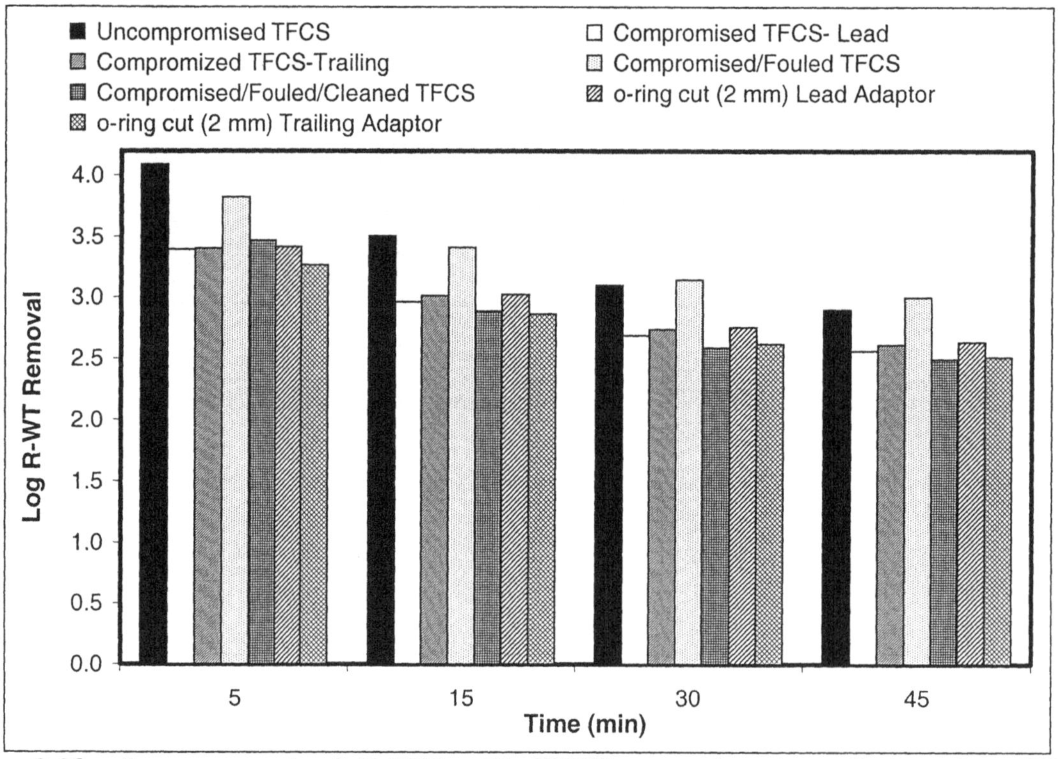

Figure 3.28 Log removal of R-WT with TFCS as a function of time for different experimental conditions in MVTU

Figure 3.29 ESPA1 RO membrane surface after fouling

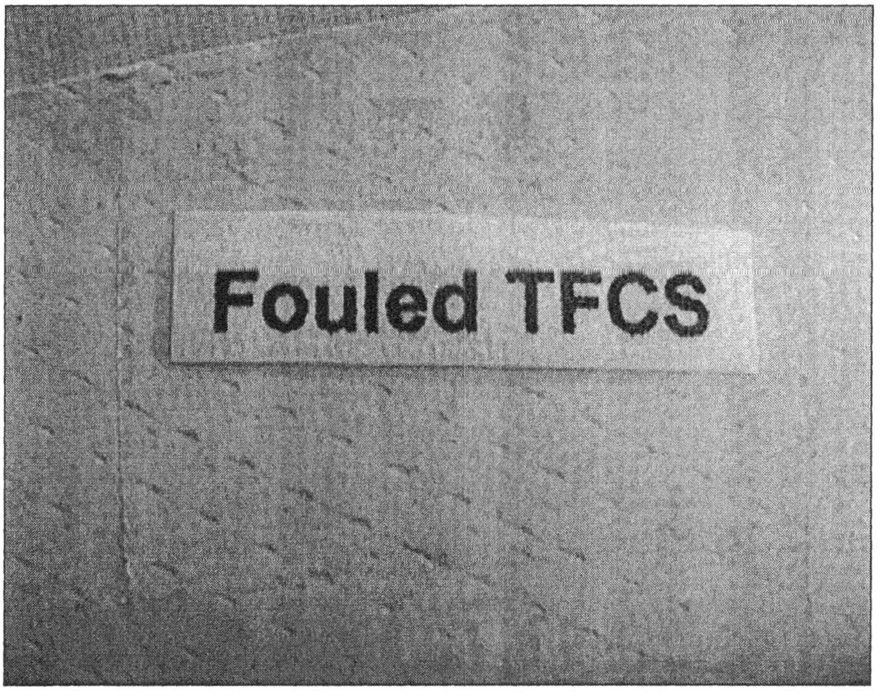

Figure 3.30 TFCS NF membrane surface after fouling

TFCS-MS2 Phage Testing. *Intact membranes and o-rings.* Log removals of MS2 *phage* as a function of time for the intact TFCS elements tested in DETU are shown in Figure 3.31. In contrast to the ESPA1 elements, which showed 6.8 log (in DETU) and 5.4 log (in MVTU) *phage* removal, all three TFCS elements in DETU showed incomplete *phage* removal, with Element#1 demonstrating only 2.8 log removal and Elements #2 and 3 about 4.3-5.3 log removal. Stated in its complement, the TFCS elements passed between 2 and 4.2 logs of MS2 *phage*, a significant degree of passage. Similarly, TFCS element used in MVTU removed only about 4-4.2 logs MS2. These results clearly indicate the presence of imperfections in the TFCS elements as fabricated. As stated earlier, the high degree of viral surrogate passage for Element#1 resulted in the need to repeat experiment W2 with TFCS Element #3. Interestingly, the log removals with the new two TFCS elements decreased somewhat with time. Figures 3.32 and 3.33 show the removal of MS2 *phage* at different experimental conditions in DETU and MVTU, respectively.

Membrane compromization. Results for the DETU compromization experiments were similar to those observed with *phage* for the ESPA1 and with the other surrogates for the TFCS. Membrane compromization significantly decreased log removals (Figure 3.32). The impact of the membrane pinhole was greatest on *phage* removal than for the other two surrogates, with removals decreasing to 1 log versus 1.5 logs for R-WT and microspheres. For MVTU, as found for R-WT, the degree of TFCS membrane compromization was not sufficient resulting in none or small reductions in log MS2 removals compared with those of intact TFCS (Figure 3.33).

Fouling and cleaning of compromised membrane. Compromised/fouled TCFS virus results exhibited the same trend as with R-WT results. For both DETU and MVTU experiments, although fouling the compromised element increased removals, the extent of increase was not as large as those found for ESPA1 element (Figures 3.32-3.33). This result again reflects the impact of the significantly smaller mass of foulant material on the surface of the TFCS membrane compared to the ESPA1 membrane. With the lower foulant accumulation, chemical cleaning was quite effective in decreasing the log removals back to levels of compromised TFCS.

O-ring compromization. Compromising the o-ring (1-mm cut) decreased the log removals in DETU by approximately 50 percent (from about 4.5-5.0 to 2.5 logs). With the MVTU, the 2-mm cut rings caused slightly greater loss of removal when installed in the lead versus the trailing position.

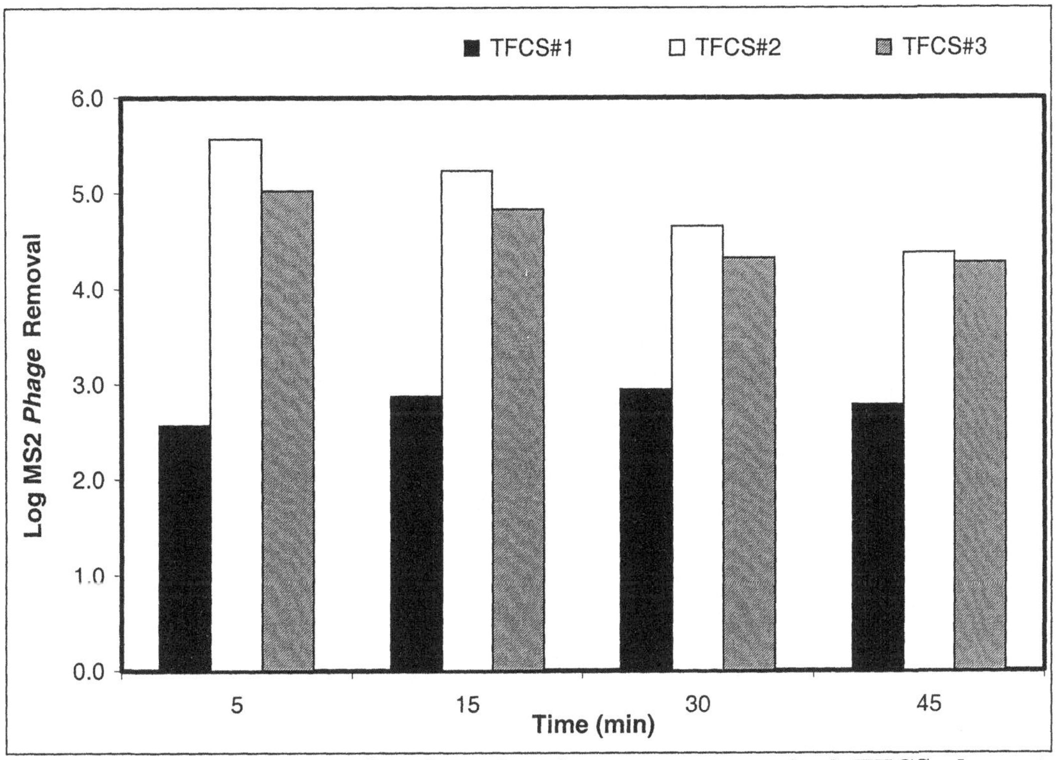

Figure 3.31 Log removal of MS2 *phage* for three uncompromised TFCS elements as a function of time in DETU

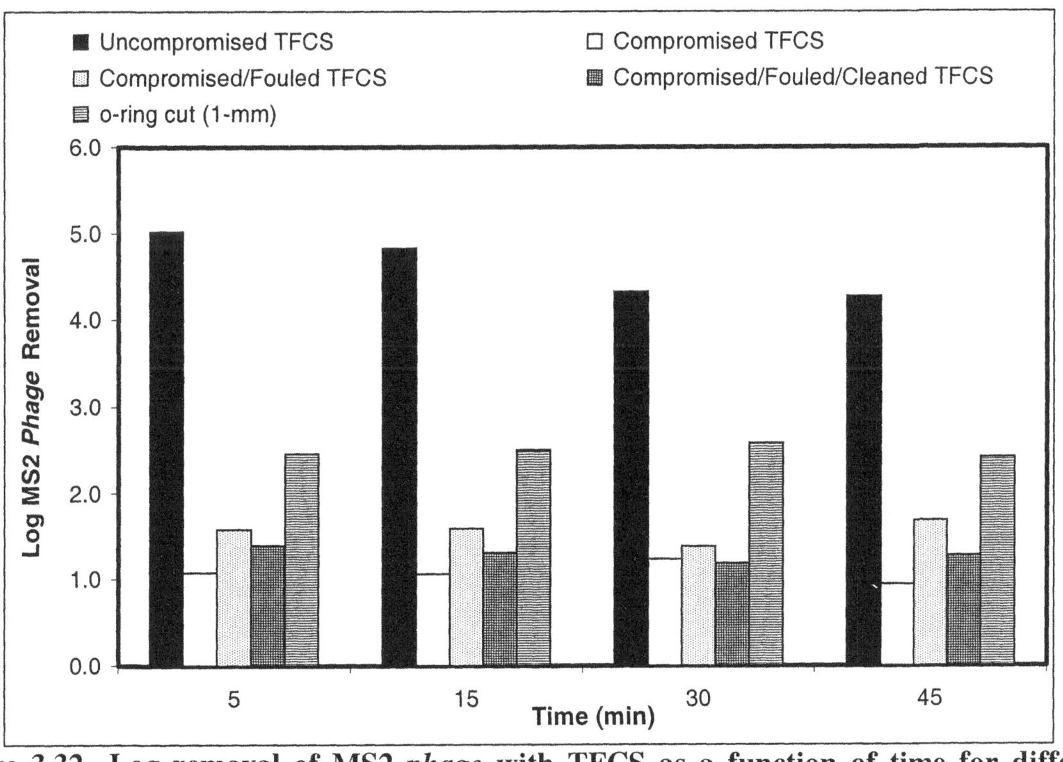

Figure 3.32 Log removal of MS2 *phage* with TFCS as a function of time for different experimental conditions in DETU

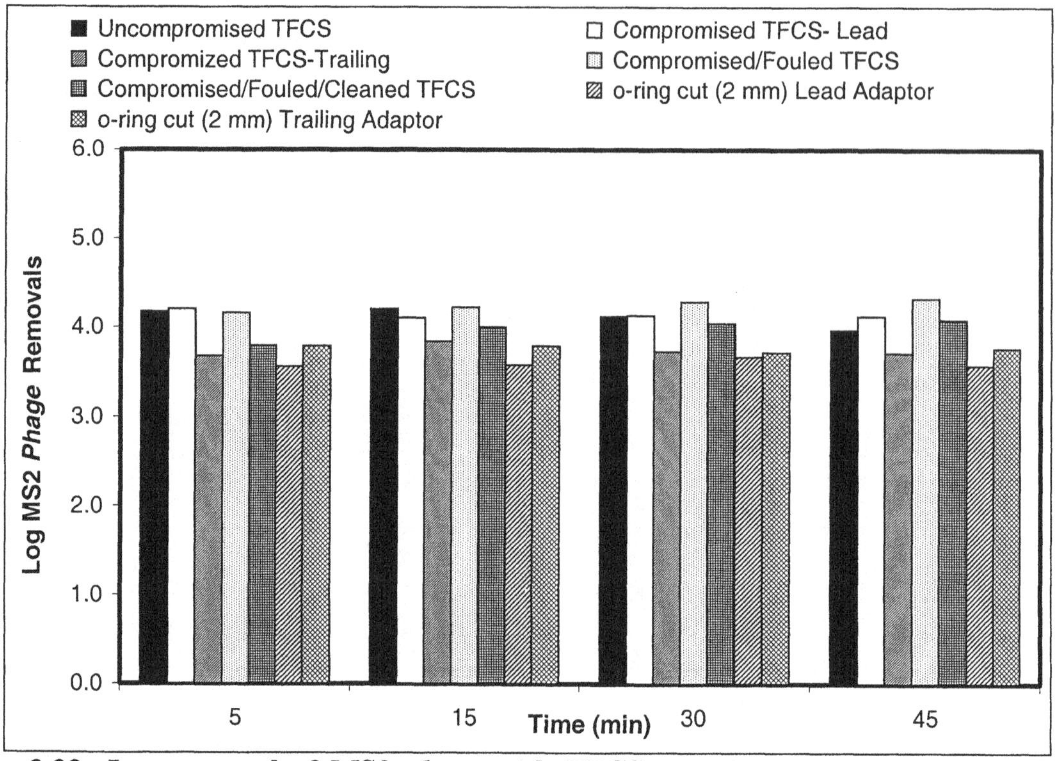

Figure 3.33 Log removal of MS2 *phage* with TFCS as a function of time for different experimental conditions in MVTU

TFCS-Microsphere Testing. Log removal of microspheres as a function of time for the TFCS elements under different experimental conditions are shown in Figures 3.34 and 3.35 for DETU and MVTU, respectively.

Intact membranes and o-rings. For DETU, log removal for intact Element (#3) ranged from 2.5-3.1, compared to 4.0 for the intact ESPA1 element. For MVTU, log removal of the intact TFCS was 4-4.3.

Membrane compromization. The impacts of membrane compromization were similar to those observed for R-WT and *phage*. Consistent with the previous results, approximately 50% reduction in log removal was observed for membrane compromization in DETU. For MVTU however, as found for R-WT and *phage*, the degree of TFCS membrane compromization was not sufficient.

Fouling and cleaning of compromised membrane. Compromised/fouled TCFS microsphere results also exhibited the same trend as with R-WT and virus results. Fouling was less effective than for ESPA1 element. Chemical cleaning off-set the gained log removals by fouling back to levels those of compromised TFCS.

O-ring compromization. There was some scatter in the data of o-ring crack experiment in DETU, possibly caused by the dynamic effect of pressure on sealing of the crack to microsphere leakage.

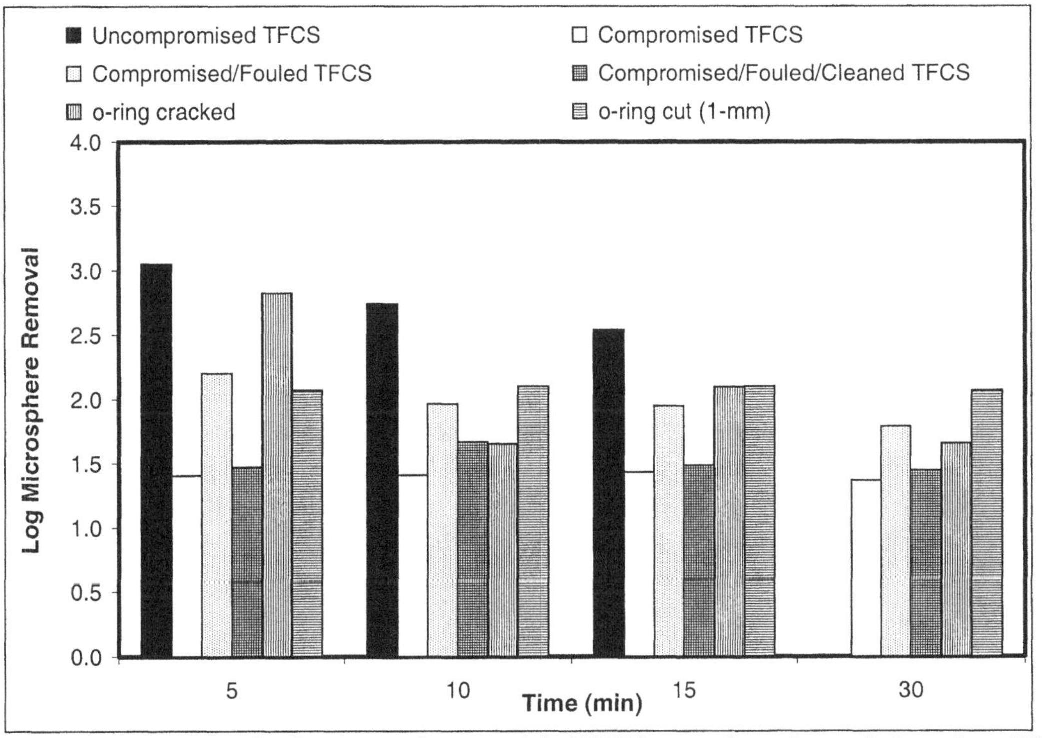

Figure 3.34 Log removal of microspheres with TFCS as a function of time for different experimental conditions in DETU

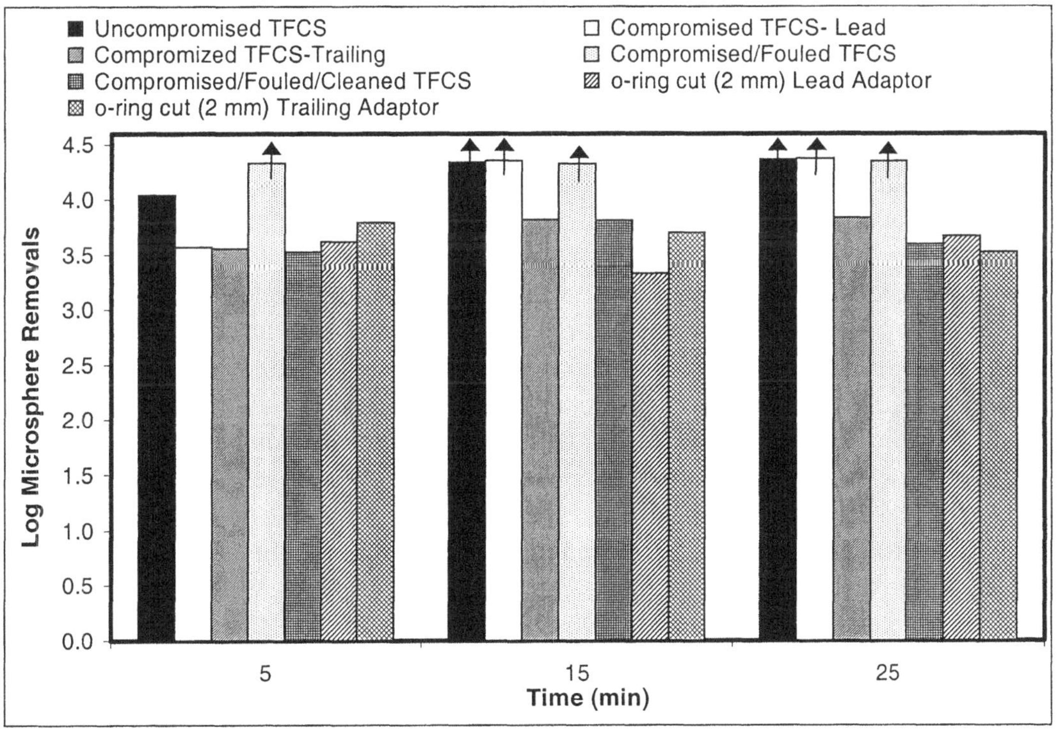

Figure 3.35 Log removal of microspheres with TFCS as a function of time for different experimental conditions in MVTU (Log removals of >4-4.3 (indicated by the upward-oriented arrow above the bar) are shown since the permeate concentrations were below quantification limit of 50 ng/L)

Uptake of R-WT

In order to determine if R-WT was being taken up by the RO/NF membranes (and possibly other test unit components), the DETU was operated for 24±2 hours following the completion of first DETU experiment in WF-21 and the amount of R-WT in the permeates measured. The results are shown in Figure 3.36. Significant uptake did occur, with the R-WT concentrations in both RO and NF permeates decreasing to at or near detection level (~20 ng/L) after 20 hours of flushing. In addition, results suggested that R-WT uptake by the membranes may vary (with the NF membrane taking up more) or that the higher concentration of R-WT present in NF permeate may require longer flushing time to be removed. In order to further test the finding that NF membrane may sorb R-WT to a greater extent than RO membrane, digital pictures of the RO and NF membrane surfaces were taken before and after dye dosing. Figure 3.37 shows the TFCS NF membrane surface after 45 min of dye dosing (part A) and after 24 hours of flushing with feed (part B). Although it is not possible to distinguish the difference in the black and white reproduction, a pinkish color (the color of R-WT) was visually seen on the NF membrane surface after dye dosing, supporting the previous finding with permeate measurements that the dye uptake by the TFCS membrane is more significant. Following 24 hours of feedwater flushing, the dye color was absent from the NF membrane. Color uptake was not observed on the ESPA1 RO membrane after dye dosing (picture not shown). These results were consistent with visual observations made of the ends of TFCS and ESPA1 elements following SETU testing. It should be noted that the active layer of both membranes is composed of the same based material, polyamide, using the same general in-situ polymerization method, although the specifics of the membrane fabrication procedure may differ from supplier to supplier, resulting in different surface roughness, charge, etc. Each manufacturer of interfacial composite polyamide membranes. To prevent the R-WT uptake from interfering with measurement of permeate R-WT concentrations, the DETU (and likewise, the MSTU and MVTU) was flushed with feedwater for a period of 24 hours prior to the initiation of the next test in the experimental matrix. Further, prior to the start of the next test, samples of the permeates were collected and analyzed to confirm the R-WT levels were below detection limit. This ensured that residual R-WT was flushed from the membrane elements (and other system surfaces) prior to dosing of microspheres in the subsequent experiment and that residual R-WT was not producing a positive interference on permeate microsphere measurements.

In the event R-WT was to be used as a viral surrogate for full-scale plant integrity testing, uptake of R-WT by membranes in the plant would have to be addressed. Current U.S EPA guidelines for the use of R-WT as a tracer in water treatment plants requires that the concentration of R-WT in drinking water not exceed 100 ng/L (Turner Designs, 2000). Given the dye uptake by the membranes, it will be essential that testing be conducted prior to full-scale integrity testing to determine what post-application flushing period is required following R-WT application before the membrane permeate can be used for drinking water. To meet the 100 ng/L requirement, the flushing period may be substantially less than the 24 hours used in this study.

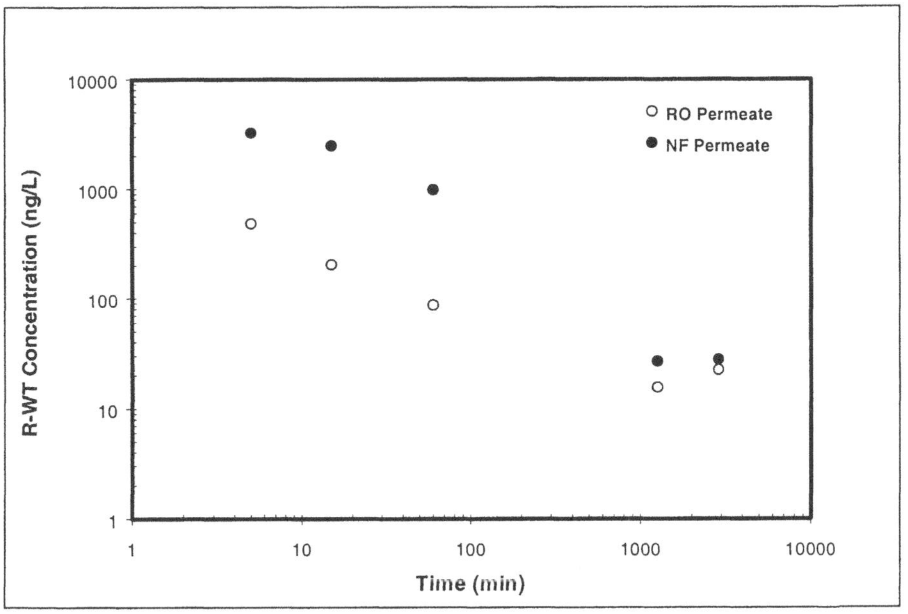

Figure 3.36 R-WT permeate concentration profile following termination of the dosing

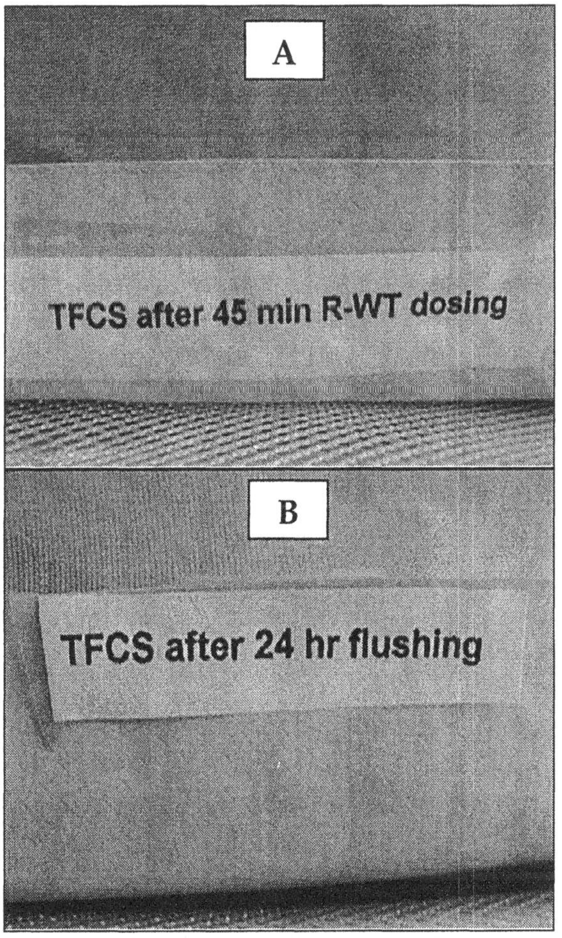

Figure 3.37 TFCS NF membrane surface after 45 min of R-WT dosing (part A) and after 24 hours of flushing with feed (part B)

Table 3.10
Surrogate removal performances of as-received ESPA1 and TFCS elements used in all test units

Surrogates	TFCS NF Log Removals			Number of TFCS elements tested
	Range[a]	Average[b]	Standard Deviation	
R-WT	2.3-4.1	2.8	0.4	5
MS2 *phage*	2.8-5.6	4.4	0.9	6
Microsphere	2.5-4.4	3.5	0.7	3
Surrogates	**ESPA1 RO Log Removals**			**Number of ESPA1 elements tested**
	Range[a]	Average[b]	Standard Deviation	
R-WT	3.7-4.1	3.7	0.6	3
MS2 *phage*	5.2-6.8	6.0	0.7	3
Microsphere	4.1-4.4	4.1	0.1	3

[a] Includes all log removals measured at different sampling times.
[b] Overall average, i.e., includes all log removals measured at different sampling times.

Performance Comparison of Intact Elements

As significant variations for R-WT and MS2 *phage* removals were found in DETU experiments between different TFCS elements received from the factory, all the ESPA1 and TFCS elements used in all pilot tests were compared for their surrogate removal performances. Table 3.10 shows the range, average and standard deviation of the log removals measured in all pilot experiments. The largest range and standard deviation was found for TFCS element in removing virus, even though maximum number of elements (6) was used in TFCS-MS2 experiments among all other cases. A MS2 removal range of 2.8-5.6 logs is an important variation, which may make consistent removal of enteric viruses at high levels (>4 logs) with this element difficult. Furthermore, one of the as-received TFCS element passed about 4 logs of MS2 *phage*, a significant degree of passage. On a general basis for all surrogates, the degree of variation in log surrogate removals between the ESPA1 elements was smaller than that found for TFCS elements. The ESPA1 RO elements also exhibited larger particulate (MS2 and microsphere) and dissolved (R-WT) surrogate removals than TFCS NF elements. The former would not be expected as the inherent "pores" in both membranes are smaller than virus-sized particles. As discussed previously, the ESPA1 elements would be expected to show improved dye removal.

Conductivity/TDS Rejection in DETU and MVTU

The percent rejection of conductivity and TDS as measured during the various DETU and MVTU experiments is presented in Tables 3.11-13. As would be expected, conductivity rejection was greater for the intact ESPA1 elements than for the comparable TFCS elements under all experimental conditions. Rejections of 98.2-98.4% were achieved by the intact ESPA1 elements during the DETU testing of all three surrogates, which were very close to the NaCl rejection of 98.6-98.9% measured during the previous SETU tests using 7 different uncompromised ESPA1 elements. Rejection by intact ESPA1 elements during MVTU tests was somewhat lower, between 95.7 and 96.6%. For the intact TFCS element (element #3) used with

DETU, conductivity rejections of 95.3-95.7% were measured, again similar to that obtained by SETU experiments, greater than the 85% and 95% advertised by the manufacturer for NaCl and $MgSO_4$, respectively (Table 3.1). As with ESPA1 results, smaller percent rejections, 88.4-90.5%, were measured during MVTU tests for the intact TFCS elements. Rejection was anticipated to be higher for DETU tests given that the ratio of divalent to monovalent ions in the MFSE is significantly higher than in the CTRW and RO, and especially NF, membranes reject divalent ions at a greater rate.

With regard to TDS (Table 3.12), intact ESPA1 and TFCS elements in DETU rejected 98.4-98.6 and 96.1-97.0%, respectively, comparable to conductivity rejection, as would be expected. This degree of rejection is <2 logs and clearly illustrate that inorganic ion rejection, as measured by either conductivity or TDS, does not provide sufficient sensitivity to be used as an accurate predictor of viral passage. This is clearly illustrated by comparing the conductivity and MS2 *phage* removals for uncompromised TFCS Elements #1 and #3 from the DETU experiments (Table 3.14). Virus removal was significantly lower for Element #1, however conductivity rejections for both elements were comparable.

Under similar conditions of compromization, conductivity and TDS rejections were consistent during the dosing of all three surrogates. This result indicates that the presence of the surrogates in the membrane feedwater at dilute concentrations did not impact the rejection characteristics of either membrane through cake formation or uptake in the rejecting layer.

Both membrane and o-ring compromization produced consistent decreases in both conductivity and TDS rejections, except in the case of the cracked o-ring. Rejections decreased as the size of the o-ring section removed increased. For example, removal of a 4-mm section decreased conductivity rejections (as measured during R-WT dosing) to 62% and 45% for ESPA1 and TFCS element, respectively. As observed with the other surrogates, the reductions in inorganic ion removal caused by membrane compromization indicate that a substantive amount of dissolved salts passed through the pinhole via advective flow. Further, and again consistent with the results for R-WT, MS2 *phage* and microspheres, deposition of a fouling layer on the surface of the compromised membranes increased inorganic ion rejection, demonstrating that such fouling, whether from colloidal organic or inorganic matter (or both) effectively plugged the pinhole induced in the membrane. The degree of plugging was more pronounced with the ESPA1 element, consistent with the degree of deposition observed and indicating that the impacts of fouling on the healing of imperfections, either from membrane casting or artificially induced, can differ by membrane supplier. Finally, cleaning of the compromised/ fouled elements with standard membrane cleaners effectively removed the foulant layer, reversing the effects of plugging on inorganic ion rejection.

Table 3.11
Percent rejection of conductivity during DETU experiments[a]

R-WT																
	Uncompromised		Compromised		o-ring cracked		o-ring cut (1-mm)		o-ring cut (2-mm)		o-ring cut (4-mm)		Compr./Fouled		Compr./Fouled/ Cleaned	
Time (min)	ESPA1	TFCS	ESPA1	TFCS	ESPA1	TFCS	ESPA1	TFCS	ESPA1	TFCS	ESPA1	TFCS	ESPA1	TFCS	ESPA1	TFCS
5	98.4	95.7	97.5	92.1	98.4	95.9	97.7	95.2	86.8	75.3	62.0	45.3	98.3	93.6	98.0	90.8
15	98.4	95.7	97.5	92.1	98.4	95.9	97.7	95.2	86.6	75.2	62.1	45.1	98.3	93.6	98.0	90.8
30	98.4	95.7	97.5	92.3	98.4	95.9	97.7	95.2	86.8	75.5	61.7	44.7	98.3	93.6	98.0	90.8
45	98.3	95.6	97.5	92.3	98.4	95.9	97.7	95.2	86.8	75.7	61.5	44.3	98.3	93.6	98.0	90.9

MS2 *Phage*																
	Uncompromised		Compromised		o-ring cracked		o-ring cut (1-mm)		o-ring cut (2-mm)		o-ring cut (4-mm)		Compr./Fouled		Compr./Fouled/ Cleaned	
Time (min)	ESPA1	TFCS	ESPA1	TFCS	ESPA1	TFCS	ESPA1	TFCS	ESPA1	TFCS	ESPA1	TFCS	ESPA1	TFCS	ESPA1	TFCS
5	98.3	95.3	97.6	91.6	98.3	95.9	97.4	94.5	NR	NR	NR	NR	98.4	93.7	97.9	90.8
15	98.3	95.3	97.6	91.6	98.3	95.9	97.4	94.5	NR	NR	NR	NR	98.4	93.7	97.9	90.9
30	98.3	95.4	97.5	91.7	98.3	95.9	97.4	94.5	NR	NR	NR	NR	98.3	93.7	97.9	90.8
45	98.4	95.4	97.5	91.8	98.3	95.9	97.4	94.5	NR	NR	NR	NR	98.4	93.7	97.8	90.9

Microsphere																
	Uncompromised		Compromised		o-ring cracked		o-ring cut (1-mm)		o-ring cut (2-mm)		o-ring cut (4-mm)		Compr./Fouled		Compr./Fouled/ Cleaned	
Time (min)	ESPA1	TFCS	ESPA1	TFCS	ESPA1	TFCS	ESPA1	TFCS	ESPA1	TFCS	ESPA1	TFCS	ESPA1	TFCS	ESPA1	TFCS
5	98.3	NR	97.4	91.3	98.3	95.7	97.4	94.7	NR	NR	NR	NR	98.4	93.7	97.8	90.6
10	98.2	NR	97.5	91.6	98.3	95.7	97.4	94.7	NR	NR	NR	NR	98.4	93.7	97.8	90.6
15	98.2	NR	97.6	91.7	98.3	95.7	97.4	94.7	NR	NR	NR	NR	98.4	93.7	97.8	90.7
30	98.2	NR	97.5	91.9	98.3	95.7	97.4	94.7	NR	NR	NR	NR	98.4	93.7	97.8	90.7

[a] Tested TFCS was element#3. NR: not run

Table 3.12
Percent rejection of TDS during DETU experiments[a]

R-WT																
	Uncompromised		Compromised		o-ring cracked		o-ring cut (1-mm)		o-ring cut (2-mm)		o-ring cut (4-mm)		Compr./Fouled		Compr./Fouled/ Cleaned	
Time (min)	ESPA1	TFCS	ESPA1	TFCS	ESPA1	TFCS	ESPA1	TFCS	ESPA1	TFCS	ESPA1	TFCS	ESPA1	TFCS	ESPA1	TFCS
5	98.6	96.1	97.7	92.9	98.5	96.3	98.0	95.7	88.0	77.0	63.9	47.3	98.5	94.2	98.1	91.6
15	98.6	96.1	97.8	92.9	98.5	96.3	98.0	95.7	87.8	76.8	64.3	47.4	98.5	94.2	98.1	91.6
30	98.6	96.1	97.7	93.1	98.5	96.3	98.0	95.7	88.0	77.2	63.8	46.5	98.5	94.2	98.1	91.6
45	98.5	96.1	97.7	93.1	98.5	96.3	98.0	95.7	88.0	77.4	63.1	46.0	98.5	94.2	98.1	91.6
MS2 *Phage*																
	Uncompromised		Compromised		o-ring cracked		o-ring cut (1-mm)		o-ring cut (2-mm)		o-ring cut (4-mm)		Compr./Fouled		Compr./Fouled/ Cleaned	
Time (min)	ESPA1	TFCS	ESPA1	TFCS	ESPA1	TFCS	ESPA1	TFCS	ESPA1	TFCS	ESPA1	TFCS	ESPA1	TFCS	ESPA1	TFCS
5	98.5	96.9	97.8	92.2	98.5	96.3	97.6	94.9	NR	NR	NR	NR	98.5	94.2	98.0	91.6
15	98.5	97.0	97.8	92.4	98.5	96.3	97.6	94.9	NR	NR	NR	NR	98.5	94.2	98.0	91.6
30	98.5	97.0	97.8	92.6	98.5	96.3	97.6	94.8	NR	NR	NR	NR	98.5	94.2	98.1	91.6
45	98.6	97.0	97.8	92.7	98.5	96.3	97.6	94.8	NR	NR	NR	NR	98.5	94.2	98.0	91.7
Microsphere																
	Uncompromised		Compromised		o-ring cracked		o-ring cut (1-mm)		o-ring cut (2-mm)		o-ring cut (4-mm)		Compr./Fouled		Compr./Fouled/ Cleaned	
Time (min)	ESPA1	TFCS	ESPA1	TFCS	ESPA1	TFCS	ESPA1	TFCS	ESPA1	TFCS	ESPA1	TFCS	ESPA1	TFCS	ESPA1	TFCS
5	98.5	NR	97.6	92.2	98.5	96.1	97.7	95.2	NR	NR	NR	NR	98.5	94.2	98.0	91.4
10	98.4	NR	97.8	92.4	98.5	96.1	97.6	95.2	NR	NR	NR	NR	98.5	94.3	98.0	91.4
15	98.4	NR	97.8	92.6	98.5	96.1	97.6	95.2	NR	NR	NR	NR	98.5	94.2	98.0	91.4
30	98.4	NR	97.7	92.7	98.5	96.1	97.6	95.2	NR	NR	NR	NR	98.5	94.3	98.0	91.4

[a] Tested TFCS was element#3. NR: not run

Table 3.13
Percent rejection of conductivity during MVTU experiments

R-WT														
	Uncompromised		Compromised-Lead		Compromised-Trailing		o-ring cut (2-mm)-Lead		o-ring cut (2-mm)-Trailing		Compr./Fouled		Compr./Fouled/Cleaned	
Time (min)	ESPA1	TFCS	ESPA1	TFCS	ESPA1	TFCS	ESPA1	TFCS	ESPA1	TFCS	ESPA1	TFCS	ESPA1	TFCS
5	95.7	88.4	95.0	87.4	94.7	83.9	95.7	87.6	94.1	84.1	96.0	88.7	93.7	84.3
15	96.0	88.5	95.1	87.4	94.7	84.0	95.7	87.7	94.2	84.3	96.0	88.8	93.9	84.7
30	95.7	88.5	95.1	87.7	94.7	84.1	95.8	87.8	94.2	84.3	96.1	89.0	94.0	84.9
45	96.1	88.9	95.1	87.6	94.7	84.0	95.8	87.9	94.2	84.3	96.1	88.9	94.1	85.1
MS2 *Phage*														
	Uncompromised		Compromised-Lead		Compromised-Trailing		o-ring cut (2-mm)-Lead		o-ring cut (2-mm)-Trailing		Compr./Fouled		Compr./Fouled/Cleaned	
Time (min)	ESPA1	TFCS	ESPA1	TFCS	ESPA1	TFCS	ESPA1	TFCS	ESPA1	TFCS	ESPA1	TFCS	ESPA1	TFCS
5	96.6	90.5	95.0	87.4	94.8	83.9	95.7	87.4	94.4	84.1	96.0	88.7	93.7	84.1
15	96.4	89.9	95.0	87.4	94.8	84.0	95.8	87.4	94.4	84.0	96.1	88.7	93.8	84.2
30	96.0	90.4	95.1	87.5	94.7	83.9	95.8	87.5	94.4	83.9	96.1	88.8	93.8	84.3
45	96.1	90.4	95.0	87.2	94.7	83.7	95.8	87.4	94.3	83.9	96.0	88.6	93.7	83.9
Microsphere														
	Uncompromised		Compromised-Lead		Compromised-Trailing		o-ring cut (2-mm)-Lead		o-ring cut (2-mm)-Trailing		Compr./Fouled		Compr./Fouled/Cleaned	
Time (min)	ESPA1	TFCS	ESPA1	TFCS	ESPA1	TFCS	ESPA1	TFCS	ESPA1	TFCS	ESPA1	TFCS	ESPA1	TFCS
5	96.3	89.7	94.7	87.1	94.6	83.7	95.9	88.0	94.3	83.6	95.9	88.4	93.3	83.6
15	96.2	90.3	94.8	87.1	94.7	83.8	95.9	87.9	94.3	83.7	95.9	88.4	93.6	83.8
25	96.4	89.9	94.8	87.0	94.7	83.7	95.9	87.7	94.3	84.2	95.9	88.4	93.6	83.9

Table 3.14
Comparison of conductivity and viral rejection by intact TFCS elements (DETU) [a]

Surrogate	Element #1	Element #3
Conductivity (%)	95.3	95.7
MS2 *phage* (log)	2.9	4.3

[a] For measurements made at 30 minutes.

MSTU Experiments

The main objective of the MSTU experiments was to investigate the impact of location of compromised element or o-ring on the removal of surrogates in a multiple stage RO system. Studies to date have focused on the integrity of bench, single element or single stage systems. These configurations do not reflect full-scale RO/NF system design, which for all but the smallest municipal plants, use either two- or three-stage arrays, with each stage having pressure vessels containing either six or seven elements. Figure 3.38 presents a simple schematic of these arrays using the minimum number of pressure vessels in each stage. The potential impact on overall integrity of compromised membrane/o-ring position is related to two factors:

- Reduction in net driving force pressure (NDF) from feed end of first stage to tail end of second stage
- Increase in concentration of contaminants, both soluble and particulate, from feed end of first stage to tail end of second stage

NDF is defined as:

$$NDF = P_f - \Delta P_{fc}/2 - \Delta\Pi - P_p$$

where:

P_f = system feed pressure, psig

ΔP_{fc} = difference in pressure between feed side of first stage and concentrate side of last stage, psi

$\Delta\Pi$ = difference in osmotic pressure across the membrane or between feed/concentrate and permeate, psi

P_p = permeate side pressure, psig

To illustrate these effects, consider a hypothetical three-stage RO system treating a 2,000 mg/L NaCl solution and having the following characteristics:

- Six-element vessels
- Ultralow pressure polyamide membrane elements having 99.4% NaCl rejection and operating at 150 psig feed pressure
- 85% recovery
- Vessel feed/concentrate differential pressure of 10 psi
- P_p of 0 psig

The NDF for the lead element in the first stage would equal:

150 psig - (2000 mg/L*0.01 psi/mg/L)[1] or 130 psig, while the NDF for the last (trailing) element in the third stage would equal:

150 psig - 30 psi/2 – [(2,000+13,200)/2 mg/L*0.01 psi/mg/L] or 59 psig (osmotic pressure term was divided by two to get an average NaCl concentration between feed and concentrate).

This difference would result in a permeate flow for the stage 3 element of only 40% of the flow of the first stage vessel. Similarly, the difference in NaCl concentration would be 13,340 versus 2,000 mg/L or a 6.6 fold concentration factor from lead to trailing positions. The net result is that as one moves the location of the membrane/o-ring imperfection from lead element position (in the first stage) to trailing element position (in the third stage), the concentration of contaminant at the imperfection (e.g., NaCl or virus) increases and the driving force for flow across the imperfection decreases. Considering the two, the impact of concentration factor is greater than the reduction in driving force.

As described in Materials and Method section (3.3) of this Chapter, the MSTU utilized a two-stage array, with each stage having two banks of vessels, each vessel holding three elements. A simplified process schematic and a picture of the MSTU are shown in Figures 3.3 and 3.4, respectively, while element locations within a vessel and vessel positions in array are shown in Figure 3.8. The MSTU was operated at 75 percent recovery for all experiments.

---

[1] The osmotic pressure of NaCl is approximately equal to the concentration times 0.01.

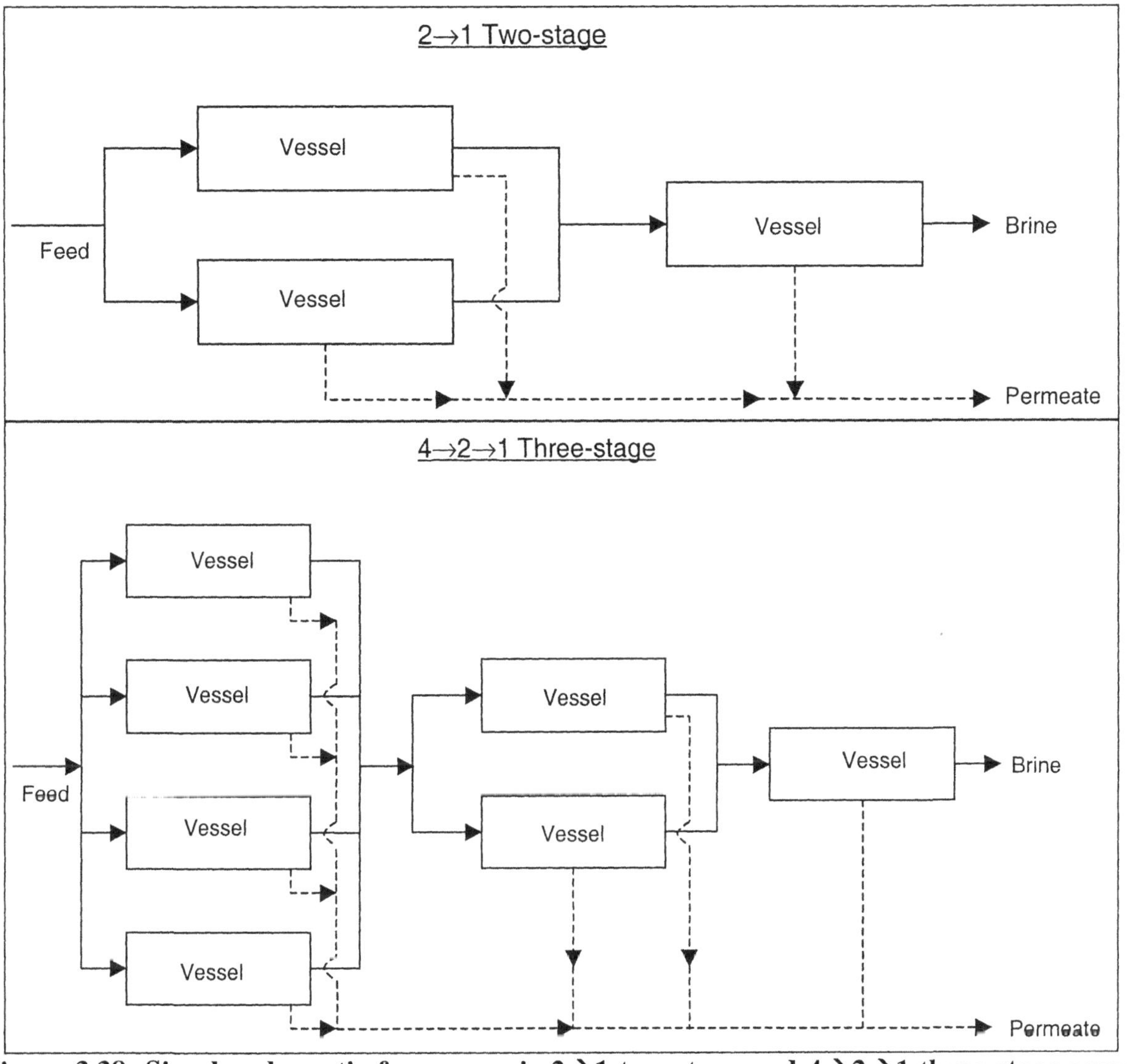

Figure 3.38 Simple schematic for a generic 2→1 two-stage and 4→2→1 three-stage array membrane system

Table 3.15 shows permeate flow contributions of each vessel to combined permeate. Flows were greatest in the first bank (vessels #1 and #2) and lowest in the last bank (vessel #6). Permeate flowrates measured volumetrically and by rotameter were in good agreement.

Table 3.15
MSTU permeate flowrates [a]

Permeate Flows (gpm)	Measured volumetrically	Measured with a rotameter[b]
P1	2.265	2.2-2.3
P2	2.305	NM[c]
P3	1.963	2.0
P4	1.945	NM
P5	1.619	1.6
P6	0.851	0.8
Sum=>	10.95	
Measured combined permeate with rotameter=>	11-11.1	

[a] Experiment conducted with all intact o-rings and membranes.
[b] Rotameter had marks with 0.20 gpm increments.
[c] Not measured.

Table 3.16
Measured pressures in MSTU [a,b]

Time (min)	Feed	Differential pressure between Bank one (vessel 1-2) feed and concentrate	Differential pressure between Bank 2 (vessel 3-4) feed and concentrate	Differential pressure between Bank 3 (vessel 5) feed and concentrate	Differential pressure Between Bank 4 (vessel 6) feed and concentrate	Differential pressure between Bank 1 feed and Bank 4 concentrate
0	143	9	5	8.5	8	30.5
15	143	9	5	8.5	7.5	30.0
30	143	9	5.5	8	7.5	30.0
60	141	9	5.5	8	7.5	30.0

[a] All pressures are in psi.
[b] Experiment conducted with all intact o-rings and membranes.

Table 3.16 shows the pressures measured for the MSTU vessels during the QA/QC test run with microfiltered secondary effluent (prior to surrogate dosing) to achieve the target flux of 10.4 gfd (11 gpm permeate flow) at 75% recovery. These pressures are representative for all MSTU experiments and did not vary appreciably with type of surrogate or imperfection. Concentrate flowrate was consistently maintained between 3.6 and 3.7 gpm. Differential pressure across each of the four banks and the total array is shown in Table 3.16.

Table 3.17 tabulates the conductivity, TDS, pH and temperature of the feed, permeates and concentrate during QA/QC test conducted before the start of compromization experiments. At the beginning of each experiment, the MSTU was operated for 60 minutes with no surrogate dosing to assess the integrity of each of the "intact" vessels (those containing uncompromised membrane elements and o-rings) by monitoring permeate conductivity, TDS, pH and temperature. For all experiments, permeate conductivity and TDS values for these vessels (which varied from experiment to experiment) were very consistent with the results of the QA/QC test, confirming integrity of the vessels. Feed conductivity and TDS values ranged around 1,450-

1,650 µS and 1,000-1,150 mg/L, respectively, during the nine MSTU experiments conducted over a period of 4 to 5 weeks. As seen from Table 3.17, conductivity was reduced to about 18-20 µs by vessel 1 and 2 in the first stage, corresponding to a removal of 98.7-98.8%. Very similarly, TDS removal was 98.8-98.9% with these vessels. Due to higher feed concentrations, vessels 3 and 4 had higher permeate conductivity and TDS concentrations than for vessels 1 and 2 and likewise, permeate conductivity/TDS for vessels 5 and 6 (second stage) were higher still. Due to the rejection of bicarbonate ion, permeate pH decreased, with the extent of the drop higher for stage 1 than stage 2. Permeate and concentrate temperatures increased compared to feed temperature because of the energy input of the high-pressure pump.

Table 3.17
Water quality parameters during QA/QC test

Conductivity (µs)									
Time (min)	Feed	P1	P2	P3	P4	P5	P6	P-composite	Concentrate
0	1565								
15	1558	20.9	18.8	27.8	29.1	47.0	68.3	33.5	5380
30	1549	20.9	18.6	27.7	28.9	46.7	68.0	33.1	5384
60	1547	20.6	18.3	27.3	28.6	46.2	67.5	32.9	5388
TDS (mg/L)									
Time (min)	Feed	P1	P2	P3	P4	P5	P6	P-composite	Concentrate
0	1120								
15	1115	13.5	12.1	18.1	18.9	30.8	44.9	21.9	4301
30	1109	13.6	12.0	18.0	18.8	30.7	44.7	21.7	4305
60	1105	13.5	11.9	17.8	18.6	30.3	44.2	21.4	4305
PH									
Time (min)	Feed	P1	P2	P3	P4	P5	P6	P-composite	Concentrate
0	7.38								
15	7.37	6.06	5.97	6.08	6.12	6.31	6.39	6.14	7.55
30	7.33	6.20	6.05	6.15	6.15	6.29	6.40	6.10	7.55
60	7.31	6.07	6.01	6.09	6.16	6.30	6.41	6.10	7.56
Temperature (°C)									
Time (min)	Feed	P1	P2	P3	P4	P5	P6	P-composite	Concentrate
0	26.9								
15	27.1	30.1	30.3	30.4	30.2	30.4	30.5	30.6	30.6
30	27.3	29.9	30.4	30.5	30.5	30.3	30.4	30.5	30.6
60	27.8	30.2	30.8	30.8	30.6	30.7	30.8	30.7	30.7

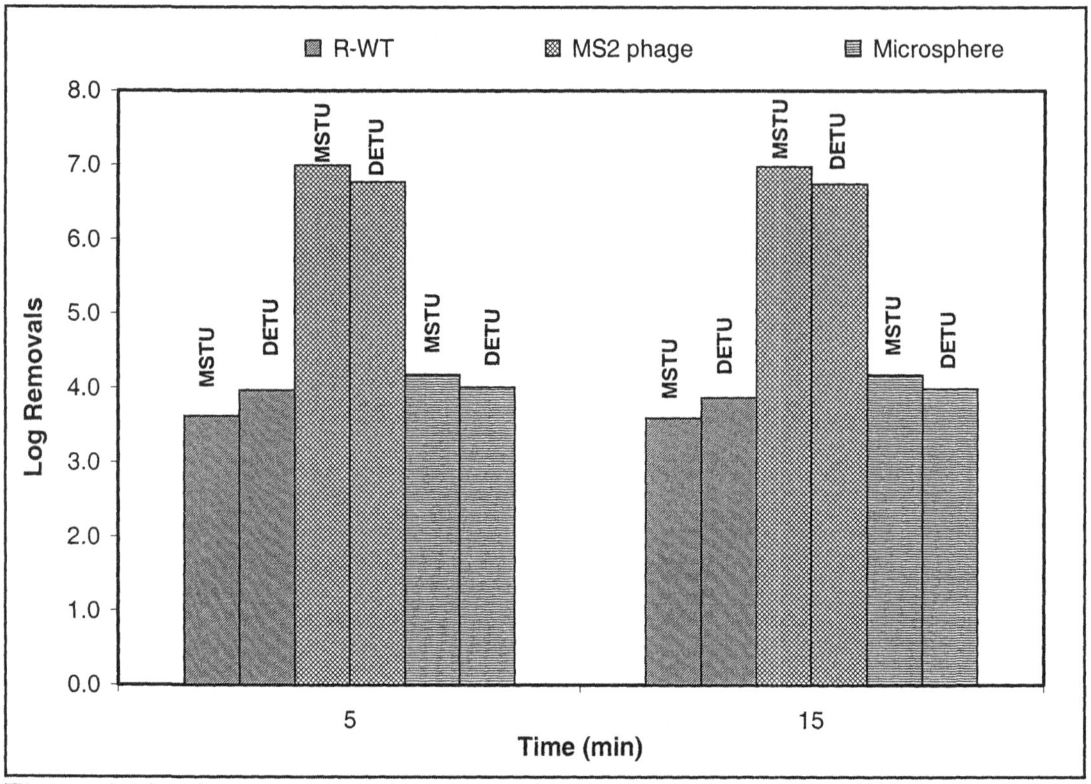

Figure 3.39 Comparison of log removals measured with DETU and MSTU (all intact system)

Surrogate Removal at High Recovery in Intact System

Figure 3.39 shows the log removals of tested surrogates (calculated based on feed and combined permeate concentrations) measured with MSTU while all elements and o-rings were intact. Log removals measured with DETU are also shown in this plot for comparison and for understanding the impact of single element vs. two-stage operation on removals. The removals of the particle surrogates (virus and spheres) in MSTU were slightly greater than for DETU while for the R-WT they were less. This may be expected considering the passage of particles through RO or NF membranes is independent of feed concentration. In contrast, for R-WT, the removal is by solution-diffusion (as is the removal mechanism for dissolved constituents) resulting in the smaller removals with MSTU due to dependence on feed concentration. This is an important difference between the surrogates and one that must be accounted for when using R-WT in full-scale systems.

The Impact of the Compromised O-ring Location on Surrogate Removals

Based on the results obtained from DETU experiments, a "2-mm cut" from the o-ring was selected and used to evaluate the impact of o-ring compromization on surrogate removal in the MSTU tests. This level of compromization provided a medium level of passage through o-ring enabling us to accurately investigate the impact of the location on the surrogate removals. The compromised o-ring was always installed in the lead adaptor of the vessel subject to compromization. As discussed in the experimental matrix (Table 3.2), vessels 1, 3, 5 or 6 were used to house the compromised o-ring to represent a different position in the two-stage array. For any given experiment, all other o-rings (and all membrane elements) were intact.

Figures 3.40, 3.41 and 3.42 show the impact of the vessel location containing the compromised o-ring on the R-WT, MS2 *phage* and microsphere log removals, respectively, relative to quality of the final combined permeate. In general, the impact for all the three surrogates was consistent with the log removals for the combined permeate decreasing as the location of the compromised o-ring was moved from bank 1 to bank 4. For example, when the compromised o-ring was installed in vessel 6 (bank 4), log removal of R-WT, MS2 *phage* and spheres was reduced from 3.7 to 2.2, 7 to 2.7 and 4.2 to 2.3, relative to the intact system. These results are consistent with theory, in that the impact of increasing feedwater surrogate concentration from bank one to bank four was greater than that of decreasing NDF pressure. Further, the lower NDF pressure (differential pressure across the o-ring) may have actually contributed to increased flow across the o-ring cut in the higher no. banks. This impact was observed in the DETU tests, where for the same compromised o-ring (2-mm cut), permeate conductivities increased with decreasing feed pressure (Table 3.18). Apparently, larger pressures provided more compression on the o-ring closing the gaps in the removed section of the o-ring. Since the operating pressures decreased in MSTU with vessel no., a negative impact on log removals would be anticipated with increasing vessel no. Thus, the combination of the two factors, decreasing NDF pressure (causing higher flow across the cut) and increasing feed concentrations, combined to produce a greater concentration of each surrogate in the combined permeate with increasing vessel (and bank) no. This result is more clearly illustrated by comparing surrogate log removals as measured for each "compromised" vessel as shown in Figures 3.43, 3.44 and 3.45 for R-WT, MS2 *phage* and spheres, respectively. With increasing vessel no, log removal decreased, with the most significant decrease occurring in the vessel 6. This trend was mirrored for inorganic ion, based on measured changes in conductivity or TDS log removals (Figures 3.46 and 3.47), however the degree of change was much less pronounced.

Table 3.18
The impact of feed pressure on permeate conductivity of ESPA1 RO element with a 2-mm cut o-ring tested in DETU

Feed Pressure (psi)	Permeate Conductivity (μs)
85	167
70	212
52	249
43	293

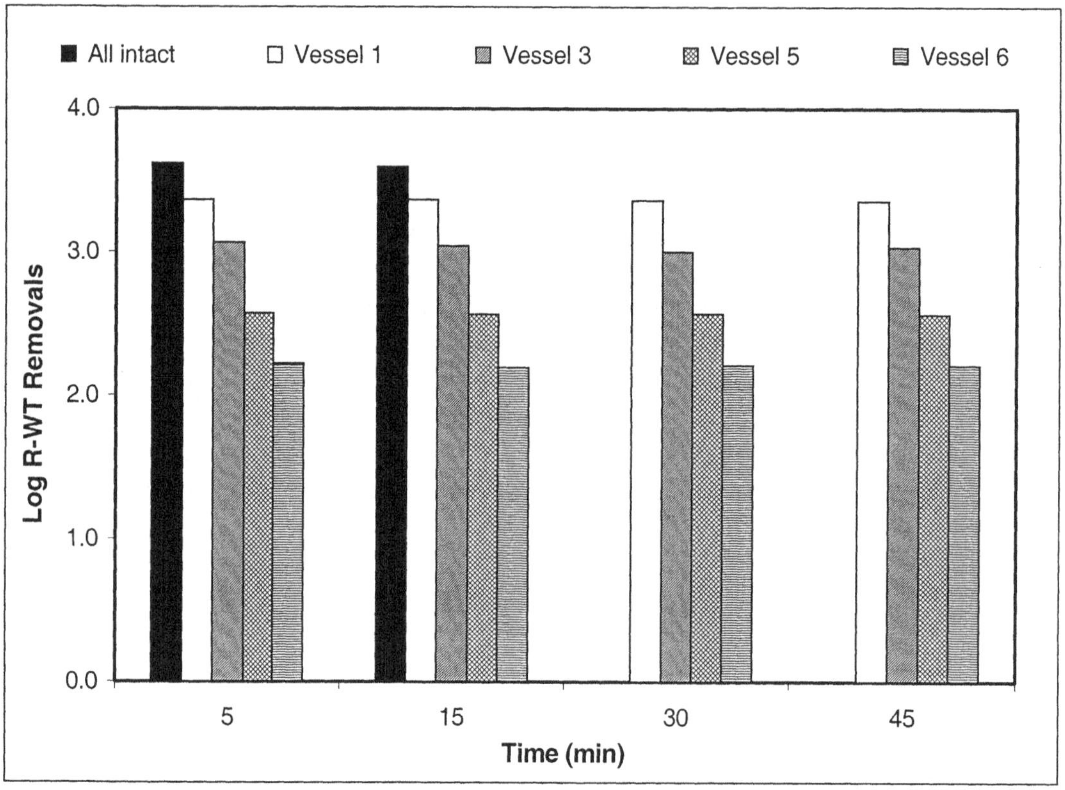

Figure 3.40 The impact of location of vessel containing the compromised o-ring on R-WT log removals with the combined permeate (5 and 15 min was the sampling time for "all intact" experiment, which also applies to the following figures)

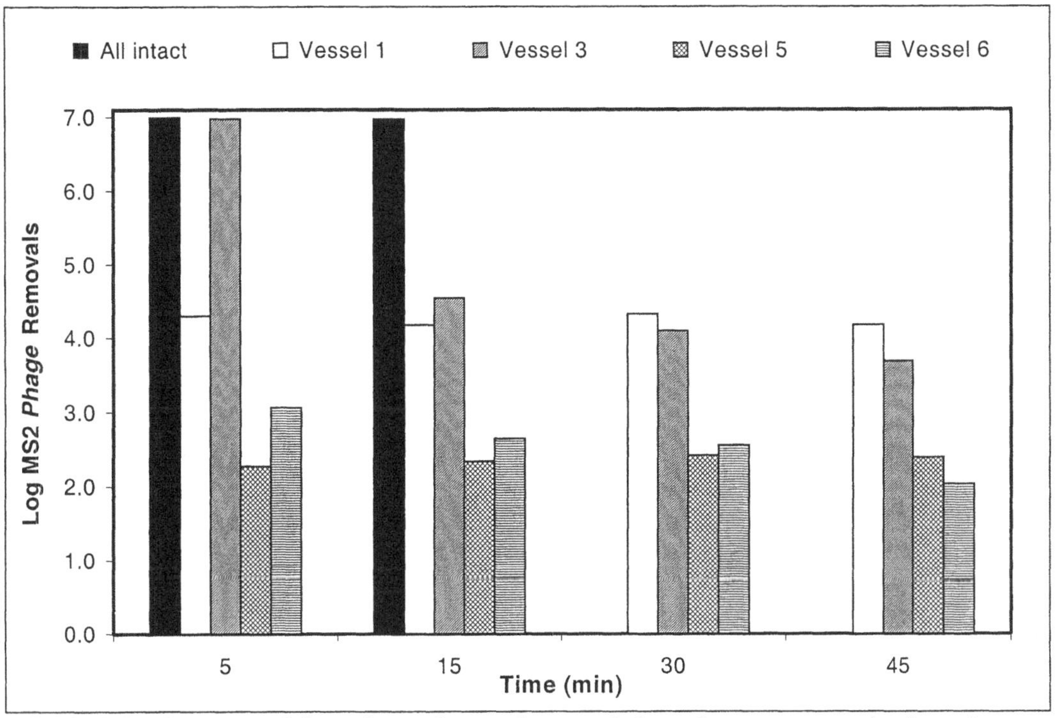

Figure 3.41 The impact of location of vessel containing the compromised o-ring on MS2 *phage* log removals with the combined permeate

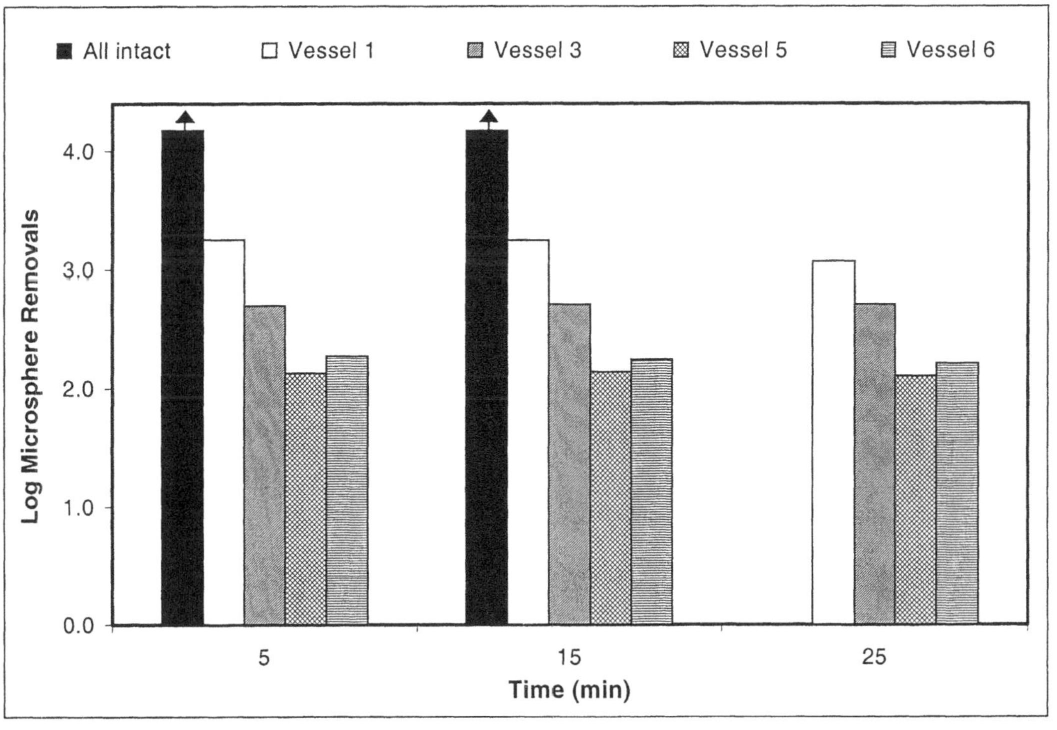

Figure 3.42 The impact of location of vessel containing the compromised o-ring on microsphere log removals with the combined permeate (Log removals of >4.2 (indicated by the upward-oriented arrow above the bar) are shown for "all intact" case since the permeate concentrations were below quantification limit of 50 ng/L)

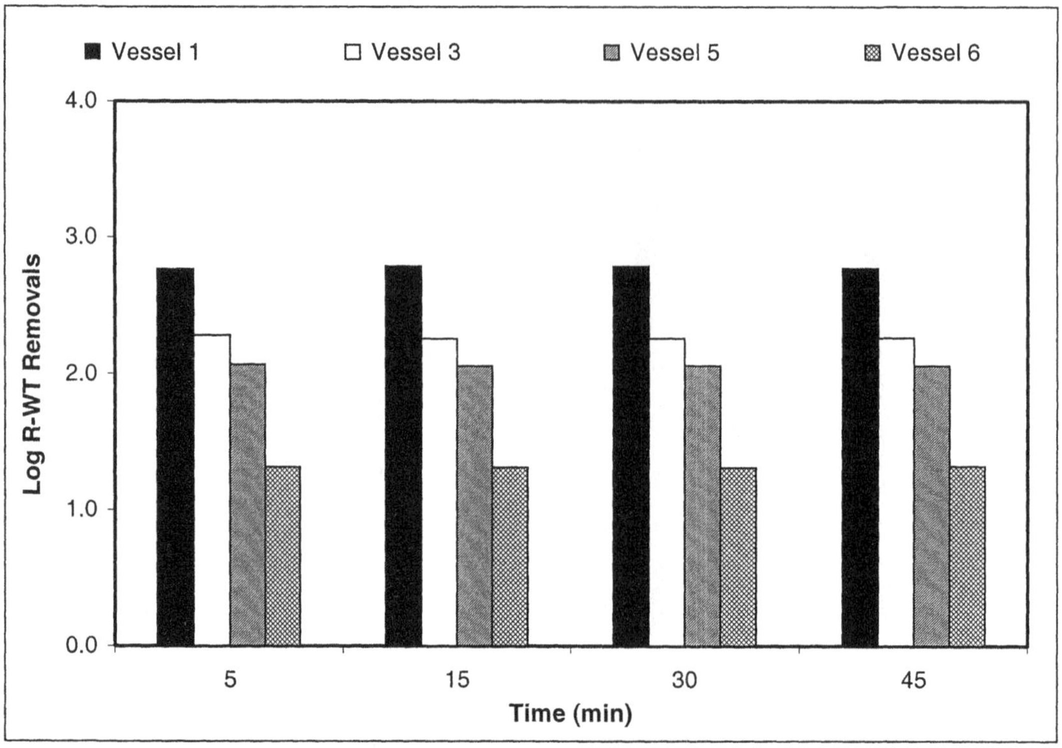

Figure 3.43 The R-WT log removals of vessels containing the compromised o-ring (Log removals for each vessel were calculated based on its feed concentration)

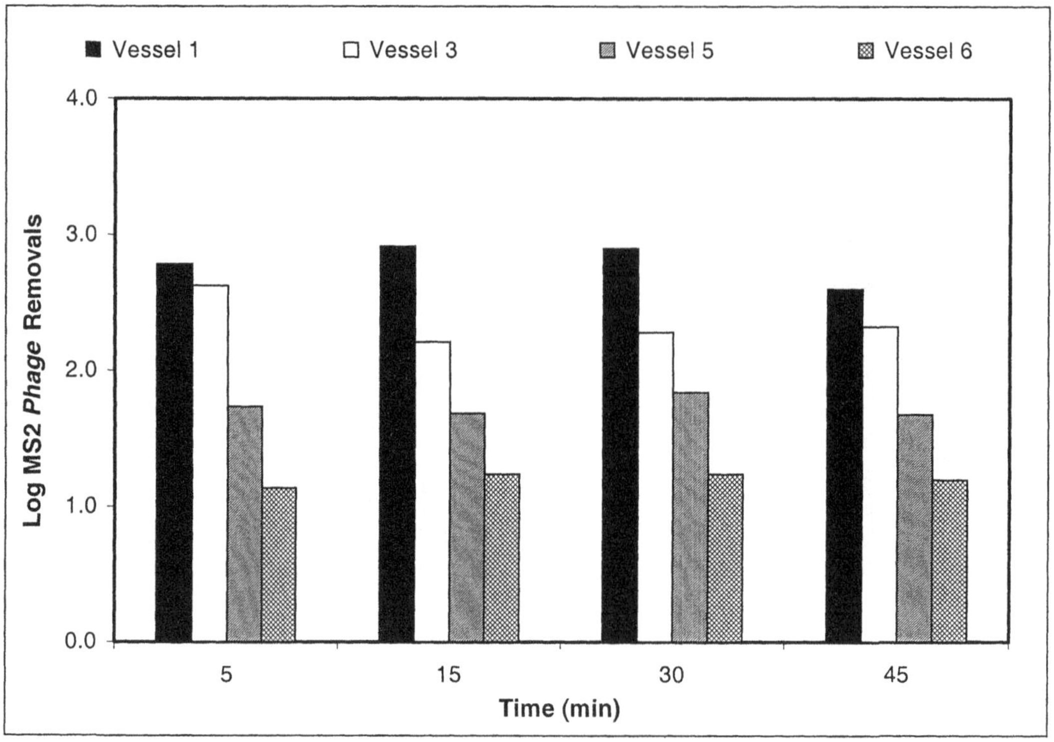

Figure 3.44 The MS2 *phage* log removals of vessels containing the compromised o-ring (Log removals for each vessel were calculated based on its feed concentration)

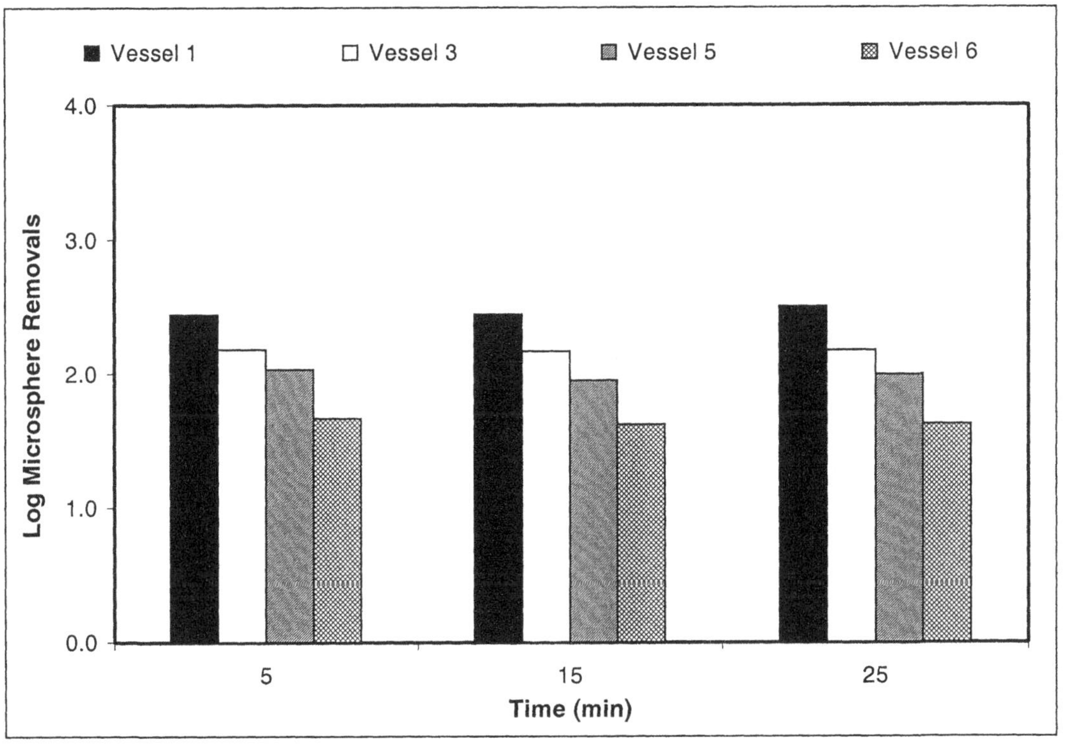

Figure 3.45 The microsphere log removals of vessels containing the compromised o-ring (Log removals for each vessel were calculated based on its feed concentration)

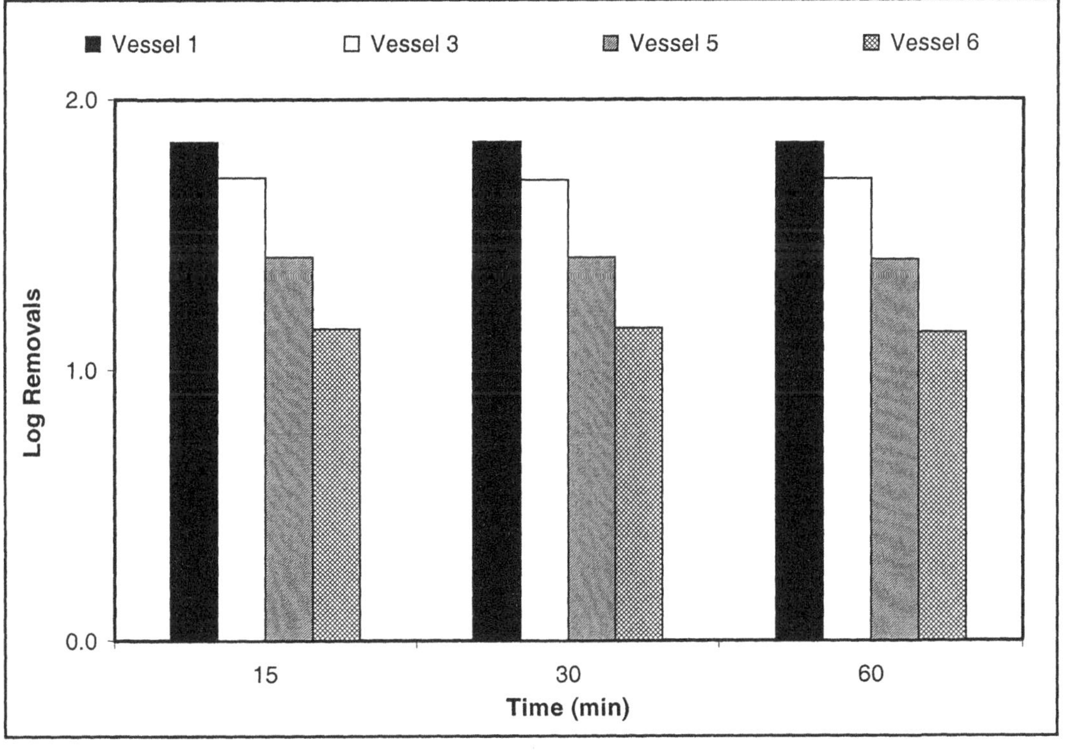

Figure 3.46 The conductivity log removals of vessels containing the compromised o-ring (Log removals for each vessel were calculated based on its feed concentration. Data obtained during initial running before any dosing)

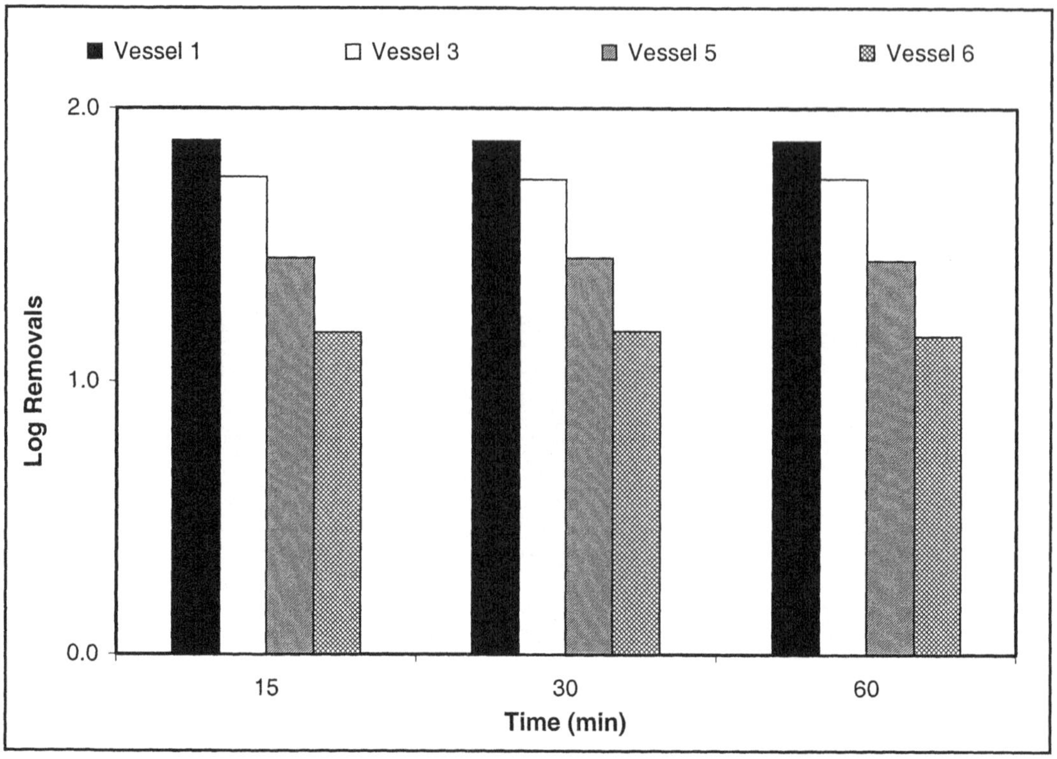

Figure 3.47 The TDS log removals of vessels containing the compromised o-ring (Log removals for each vessel were calculated based on its feed concentration. Data obtained during initial running before any dosing)

The Impact of the Compromised Membrane Location on Surrogate Removals

Membrane compromization was performed with the same technique as used in DETU and MVTU tests. A pinhole (approximately 300-500 μm in diameter) was induced in one side of an ESPA1 RO membrane leaf. This level of compromization provided a medium level of surrogate passage. The element containing the compromised membrane was installed as the lead element in each vessel subject to testing. As shown in Table 3.2, and like the o-ring experiments, vessels 1, 3, 5 or 6 contained the compromised element. Except for the one compromised element used in a particular experiment, all other (17) RO elements and o-rings were intact.

Figures 3.48, 3.49 and 3.50 show the impact of the vessel location containing the compromised element on the R-WT, MS2 *phage* and microsphere log removals, respectively, relative to the quality of the combined permeate. While the trends for o-ring experiment results were consistent for all surrogates, significant differences exist in the results for the compromised membrane tests depending on whether the surrogate is particulate or dissolved. For MS2 *phage* and microspheres (Figures 3.49 and 3.50), the log removals for the combined permeate showed an increase with increasing vessel no (The exception is the MS2 *phage* data for vessel 1 compromised experiment). This trend suggests that for the particle-based surrogates the rate of passage of the particles through the membrane imperfection is controlled more by NDF pressure than by feed concentration. Unlike the o-ring, one would not anticipate that differences in NDF pressure would impact the size of the imperfection, however, it is possible that a higher concentration of particles could more effectively occlude the imperfection, thereby reducing the rate of advective flow through the pinprick.

For R-WT, log removal for combined permeate was essentially constant, again except for R-WT data of vessel 1 (Figure 3.48). As R-WT is a dissolved organic compound, its rate of passage would increase with increasing feed concentration, thereby decreasing removal even in the absence of a membrane imperfection. This, combined with advective flow through the imperfection, would result in higher concentrations of R-WT in the permeates of the higher no. vessels compared with *phage* and spheres. Further, the R-WT would not be expected to cause occlusion and reduce advective flow. The solute behavior of the R-WT is corroborated by the trends in log removal for conductivity and TDS (Figures 3.51 and 3.52, respectively).

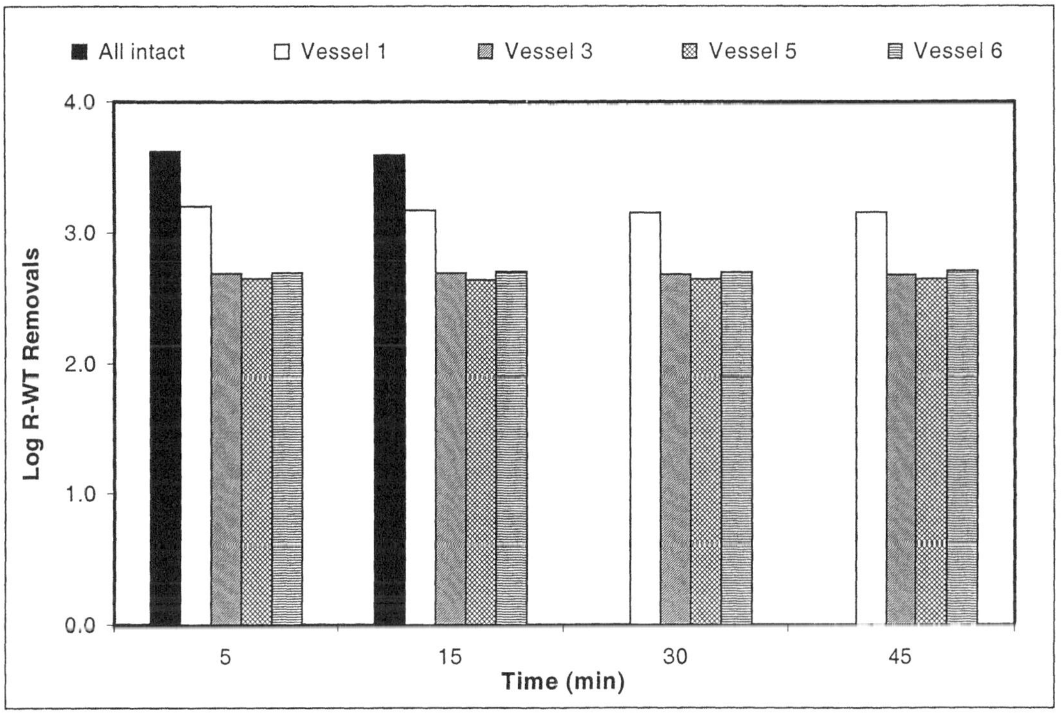

Figure 3.48 The impact of location of vessel containing the compromised element on R-WT log removals with the combined permeate (5 and 15 min was the sampling time for "all intact" experiment, which also applies to the following figures)

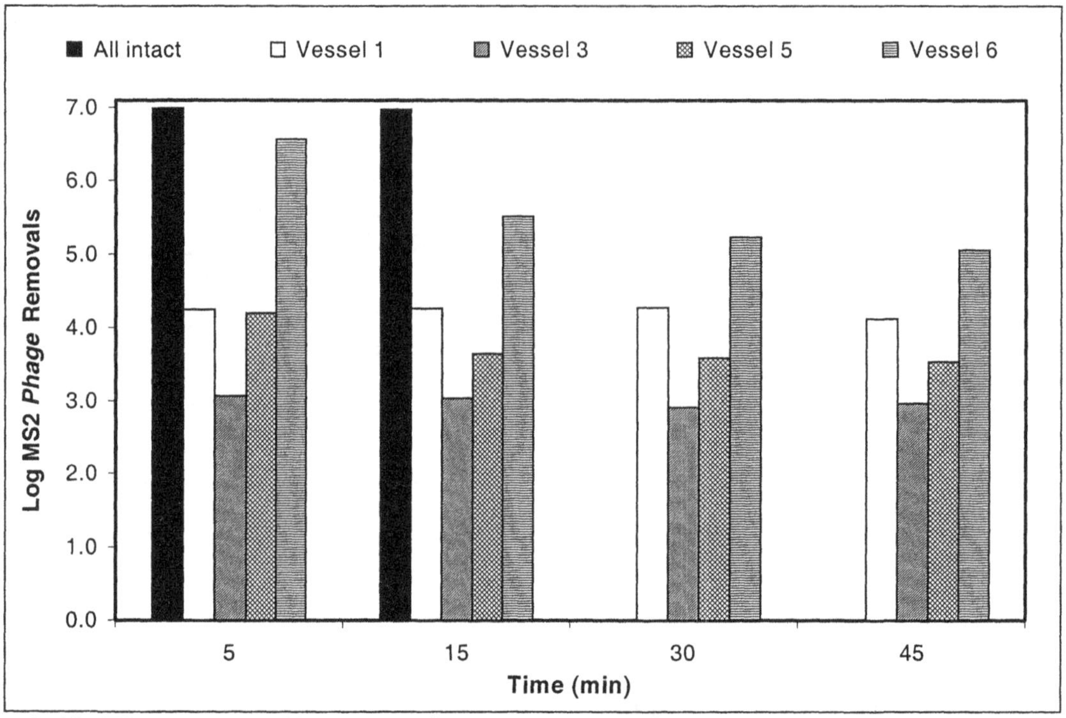

Figure 3.49 The impact of location of vessel containing the compromised element on MS2 *phage* log removals with the combined permeate

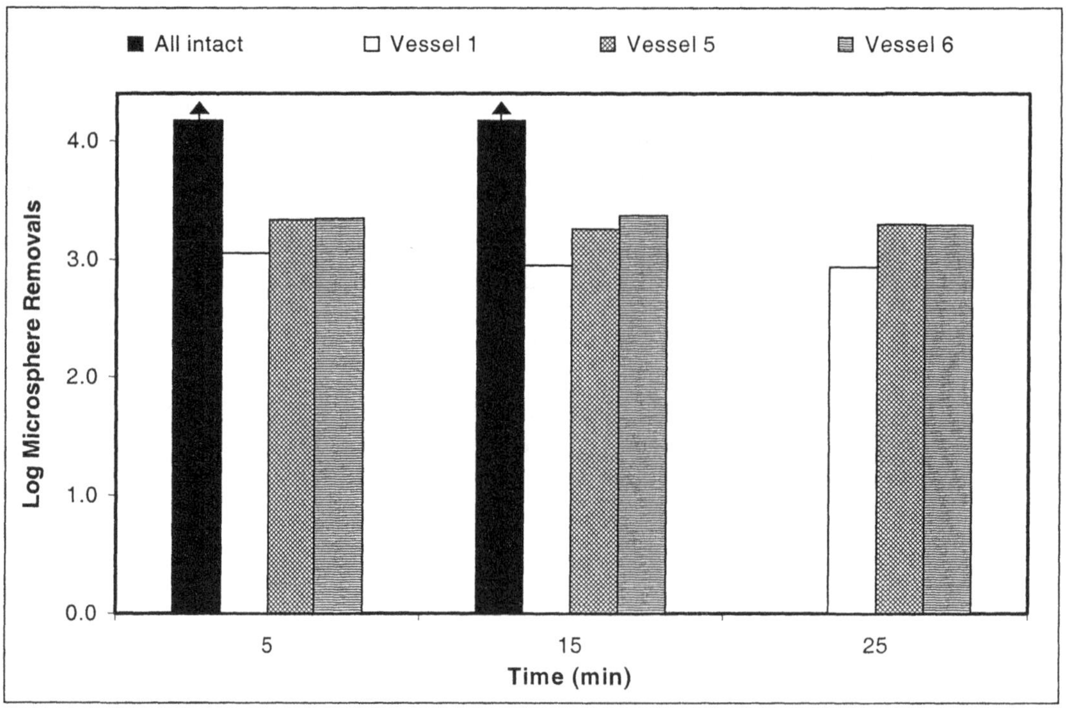

Figure 3.50 The impact of location of vessel containing the compromised element on microsphere log removals with the combined permeate (Log removals of >4.2 (indicated by the upward-oriented arrow above the bar) are shown for "all intact" case since the permeate concentrations were below quantification limit of 50 ng/L)

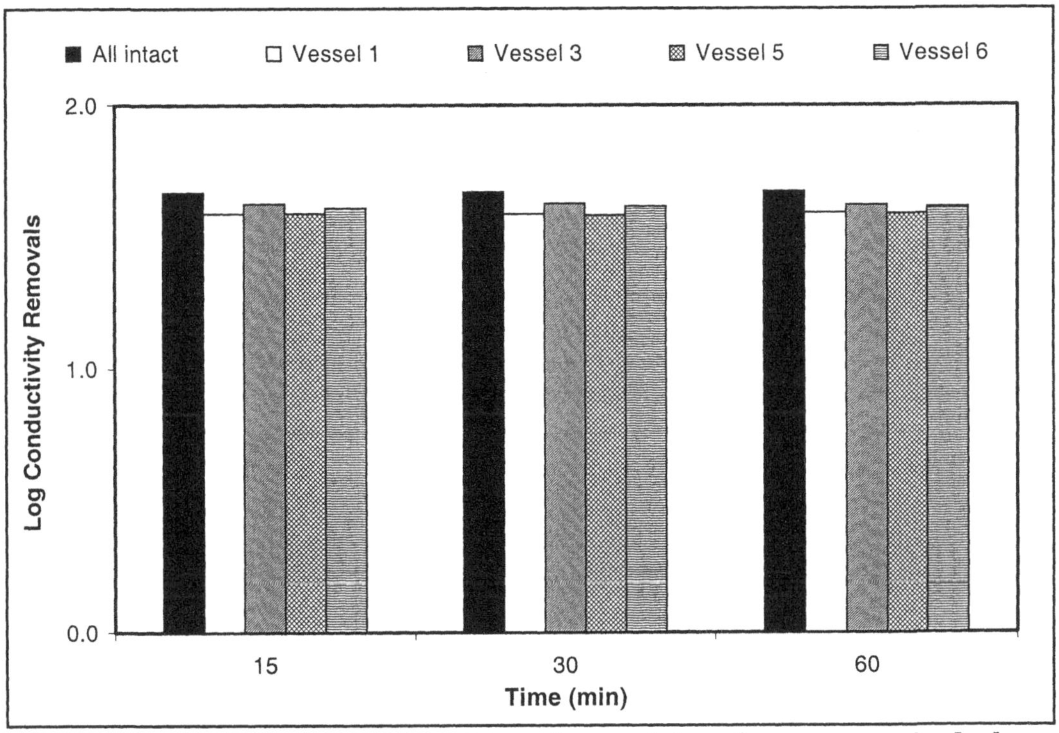

Figure 3.51 The impact of location of vessel containing the compromised element on conductivity log removals with the combined permeate.

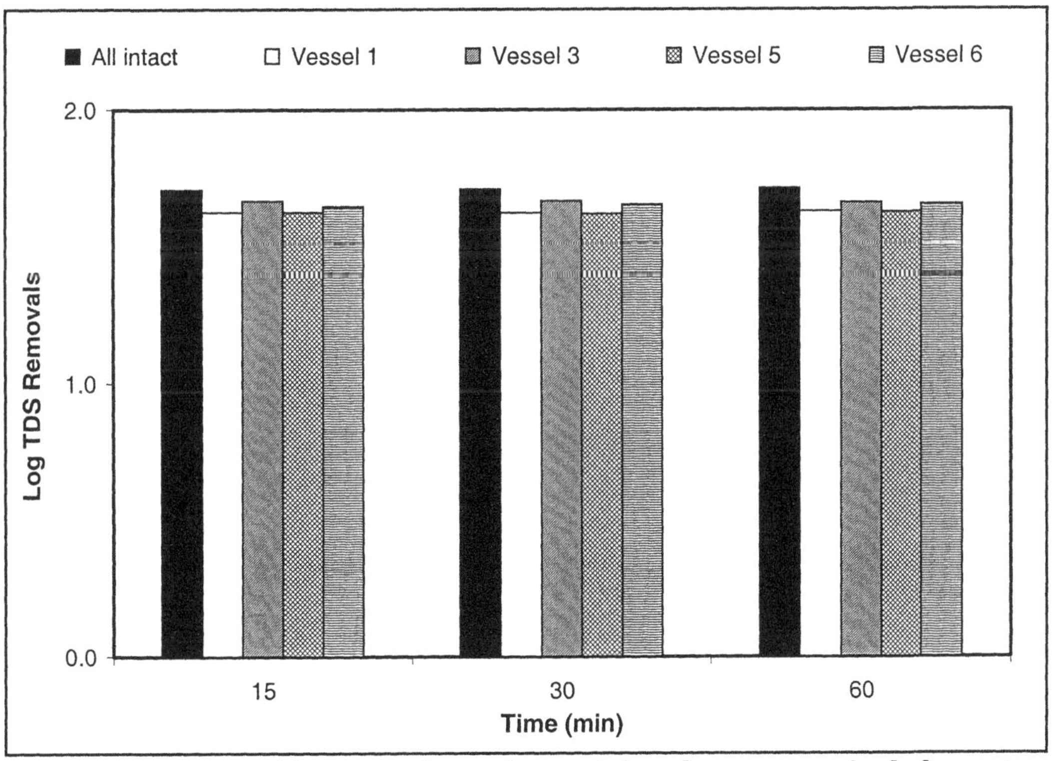

Figure 3.52 The impact of location of vessel containing the compromised element on TDS log removals with the combined permeate

Comparison of Tested Surrogates for Full-Scale Integrity Testing

Results obtained in this project showed that conductivity and/or TDS rejection cannot be used as an accurate predictor of viral passage in high-pressure membranes. Intrinsic imperfections are not distinguishable using normal manufacturer's QA/QC "wet testing" which simply measures percent rejection of a salt solution. Given the minimum quantification level of the spectrofluorometer used in this research (about 50 ng/L), a microsphere challenge level of about 4-4.5 log could be achieved using a cost-feasible feed concentration of 1 mg/L at pilot scale. However, the use of microspheres of the type used in this research as a viral surrogate in full-scale applications is limited by microsphere cost (as discussed in the following sub-section) as well as the requirement of sophisticated instrumentation for its analysis. For MS2 *phage* as a biological surrogate, challenge concentrations of >7 logs could be readily achieved with the *phage* culturing techniques developed in this project. However, the cost and effort to culture and plate sufficient quantities of MS2 phage for full-scale application will be burdensome.

A new non-biological surrogate, Rhodamine WT, may be a practical surrogate for detecting imperfections in both RO and NF membranes relative to virus removal. It is inexpensive, easy to dose, and can be accurately measured at concentrations down to 20 ng/L in membrane permeates, resulting in an ability to measure about 4 log removal at a feed concentration of <2 mg/L. However, R-WT adsorbs in the membrane (with the degree of adsorption dependent on membrane type), requiring that flushing must be performed to remove its presence in the permeate following application. Removal of R-WT by the RO/NF membrane is by solution-diffusion and dependent on feed concentration, which should be accounted for when comparing results between single-element test units and multi-stage systems as in full-scale applications. Therefore, use of R-WT as a surrogate may not be ideal for all applications due to inherent differences in the mechanism by which dye and virus are removed and the limit of rejection (approximatly 3 logs for NF and 4 logs for RO).

For all tested surrogates, the presence of low concentrations during dosing did not cause membrane fouling or modification to the conductivity/TDS rejection characteristics either through cake layer formation (*phage* or microspheres) or adsorption within the membranes (R-WT). This indicates that use of such surrogates at full-scale would not be constrained by fouling considerations.

Economic Comparison of Non-Biologic Surrogates

Costs for use of microspheres and R-WT for full-scale system integrity testing were estimated baed on costs realized for pilot-scale application. Based on a feed concentration of 1 mg/L and unit chemical cost of $0.50 per mg of microspheres, the "unit use" cost for integrity testing in a membrane system is estimated at $1.90 per gpm of feed flow per minute of application. For a 5-mgd (permeate flow) plant operating at 85% recovery, in which the microspheres are dosed for a period of 10 minutes, this equates to a microsphere cost of $82,46 per integrity test[2]. In contrast, at the same feed concentration of 1 mg/L and unit chemical cost of $0.0037 per mg R-WT, the cost for an integrity test using R-WT is approximately $0.01 per gpm feed flowrate or about 0.5 percent of the cost of microspheres. This results in a full-scale cost of

---

[2] Calculated as follows: $1.90/gpm-minute x 4340 gpm x 10 minutes = $82,465.

only $434 for the same plant capacity, recovery and use duration[3]. Based on these costs, integrity testing with R-WT is economically practical using today's commercially available product.

The microsphere product employed in this research is manufactured for "high end" biomedical uses, where small quantities are employed for specialized applications. To make microspheres applicable to municipal high pressure membrane market where much larger quantities are needed, a microsphere product would have to be produced at 1/100th the cost and require less expensive analytical instrumentation and methods for measurement. Very preliminary discussions were held with one chemical supplier, who indicated that microspheres could be manufactured with these characteristics if the market could support their use (Moll, 2002).

---

[3] Calculated as: $0.01gpm-min x 4340 gpm x 10 minutes = $434.

CHAPTER 4
SUMMARY AND CONCLUSIONS

The primary conclusions drawn from this project are summarized below in subsections.

EXPERIMENTAL METHODOLOGY

- Uptake of microspheres and MS2 *phage* by system components experienced with the closed loop SETU testing was not observed with the flow through DETU/MSTU/MVTU experiments. Furthermore, the feed R-WT, MS2 *phage* and microsphere concentrations were very stable during the 45-minute dosing with percent relative standard deviations less than 19% (the majority around 3-7%) between the measured feed concentrations. This indicates that each surrogate was spiked at consistent levels in DETU, MVTU and MSTU.

UNCOMPROMISED (INTACT) ELEMENTS

- At the conditions utilized in this research, intact ESPA1 RO elements, as received from the manufacturer, tested both in flat-leaf and spiral-wound configurations, demonstrated >6 logs removal of MS2 *phage*; 4-log removal of R-WT and 4-log removal of microspheres (the maximum removal that could be achieved at the concentrations applied). Intact TFCS NF elements showed lower surrogate removal: 2.5 to 4 logs for R-WT and microspheres and between 3 and 5.5 logs for MS2 *phage*. The lower removal for MS2 *phage* and microspheres were not anticipated for the NF membrane, whose particle rejection should be complete, like RO, due to the non-porous nature of high pressure membranes. Reduced R-WT rejection by NF, relative to RO, could be expected based on the larger MWCO of the NF membrane. However, the NaCl and/or conductivity/TDS rejections of the as-received ESPA1 RO and TFCS NF elements tested were found to be within or more than the manufacturer's specifications.
- The largest range and standard deviation in log removals with TFCS was for MS2 *phage*. One element passed approximately 4 logs of MS2 *phage*, a significant degree of passage. In contrast, the variability in surrogate removals by ESPA1 was smaller compared to those by TFCS.
- Conductivity rejection by both RO and NF decreased with increasing recovery (i.e., was higher with DETU than MVTU tests) and was greater with MFSE than CTRW. Both results are consistent with the solution-diffusion model for thin film membrane behavior.
- Surrogate removal by both RO and NF was not affected by source water type.

EFFECT OF MEMBRANE COMPROMIZATION

- Of the various chemical and physical approaches used to compromise membrane integrity (limited exposure to free chlorine, air entrainment, scratch, and pinholes of various sizes) only pinholes fully penetrating both the active rejecting and support layer of a membrane leaf produced significant increases in the passage of both biological and non-biological viral surrogates as well as conductivity and TDS.

- In the bench-scale coupon tests, flat-sheet membrane coupons compromised by inducing 10-μm pinholes penetrating both the surface (active) and support layers reduced *phage* rejection to 2-5-3.0 logs. Similar log reductions were observed with the microspheres. Pinholes of 100-μm reduced removals to 1-1.5 logs of both *phage* and spheres. Virus passage through induced imperfections decreased to non-detect after 24 hours of operation. This phenomenon suggests that imperfections are plugged by particles (i.e., low levels of corrosion-induced particles) in the feedwater..
- For the pilot-scale experiments, membrane compromization also significantly decreased log removals of all surrogates, as well as conductivity and TDS. The similar degree of removal reductions for all surrogates (about 50%) suggest that the pinhole (approximately 300-500 μm in diameter) produced advective flow that allowed passage of both particulate and dissolved contaminants independent of surrogate type and physicochemical characteristics. This further suggests that the passages of enteric viruses through the membrane imperfections (i.e., a pinhole) may be satisfactorily predicted by both the particulate and dissolved surrogates used in this study, as long as the size of a pinhole is sufficiently large not to cause a steric hindrance to passage of viral size particles. This is an important finding in that a dissolved surrogate (R-WT) with its attendant advantages of low cost and ease of use has potential to be a practical method for assessing viral integrity of high pressure membranes at full-scale.
- Since the method of membrane compromization used in this research employed hand technique, some variations in the performance of compromised membranes was observed.

EFFECT OF O-RING COMPROMIZATION

- A crack induced along the inner surface of the o-ring was insufficient to cause a loss of system integrity. Removing small sections from o-rings did impair integrity. Log removals decreased proportionately with increasing size of removed section and were very sensitive to the degree of o-ring compromization.

EFFECT OF THE LOCATION OF THE COMPROMISED MEMBRANE OR O-RING

- For MSTU testing, log removal of all three surrogates, as well as conductivity and TDS, as measured in the combined permeate, decreased as compromised o-ring location was moved from front to tail end of the two-stage array. The combination of two factors, decreasing NDF pressures (causing higher flow across the cut) and increasing feed concentrations, could explain this result. Larger NDF pressures appear to provide more compression on the o-ring closing the gaps in the removed section of the o-ring thus decreasing the passage to permeate, as also proved during DETU testing. Increasing feed concentration increases the mass of surrogate present in the flow through the imperfection.

- The effect of array location on surrogate log removal measured in the combined permeate by a single membrane-compromised element varied depending on the type of surrogate. Log removals for the particulate surrogates (*phage* and spheres) increased as the compromised element was moved from vessel 1 to vessel 6; for the soluble R-WT, the impact of vessel location was negligible. This suggests that for the particle-based surrogates (removed by "physical sieving" mechanism) the rate of passage of the particles through the membrane imperfection is controlled more by NDF pressure than by feed concentration. It is also possible that a higher concentration of particles could more effectively occlude the imperfection, thereby reducing the rate of advective flow through the pinprick. Dissolved R-WT, removed by "solution-diffusion" mechanism, would not be expected to cause occlusion and reduce advective flow.
- The results for the impact of compromised membrane or o-ring location in the three-element vessel (MVTU) on removals were not conclusive. On a general basis, trailing-positioned compromised o-ring or membrane exhibited larger log removals than those of lead-positioned ones. While this is difficult to explain based on insignificant operating pressure differences between the lead and trailing elements or o-rings, experimental errors or the hand variations in compromization technique may be important factors.

EFFECT OF FOULING OF COMPROMISED MEMBRANE

- For both the MFSE and CTRW, fouling of compromised ESPA1 elements increased the log removals of all surrogates (including conductivity and TDS) to levels of intact elements, even somewhat higher in some cases. This indicates that fouling plugs the hand-induced imperfection as well as those induced during manufacture. Although fouling also increase removals of compromised TFCS NF elements, the extent of increase was not as large as with ESPA1. The difference was attributed to the differing degree of membrane fouling and indicates that the fouling properties of the two thin film polyamide membranes differ. Bench-scale testing also showed improvements in log removal by ESPA1 from corrosion-product induced fouling. These results overall suggest that fouling, presumably through the bridging of holes in the membrane layer by feedwater particles, can "heal" minor imperfections present from manufacture as well as the larger imperfections induced in this research, however the degree of healing will be proportional to the membrane's propensity for fouling.

EFFECT OF CHEMICALLY CLEANING OF FOULED/COMPROMISED MEMBRANE

- Chemically cleaning the compromised and fouled elements decreased the log removals of surrogates, conductivity and TDS, showing that the increases in log removal by fouling is reversible and that the mechanism for increased removal is plugging of the imperfections (in contrast to a change in the morphology of the membrane by compaction). The degree of reversibility was a function of the nature of the foulant layer. Cleaning was more effective when performed on elements fouled by operation on MFSE (DETU testing) than by operation on CTRW (MVTU testing).

This result may be due to either differences in the cleaning mixtures used (both were alkaline based but used different compounds) or the physicochemical nature of the cake layers formed by treating MFSE versus CTRW. The trends observed in this project regarding the impacts of fouling and chemical cleaning on virus removal are similar to those found for low-pressure membranes (MF and UF) in other studies (Jacangelo *et al.*, 1995).

COMPARISON OF TESTED SURROGATES

- Virus challenge concentrations of >7 logs could be readily achieved with the *phage* culturing techniques developed in this project. Given the minimum quantification level of the spectrofluorometer used in this research (about 50 ng/L), microsphere challenge level of about 4-4.5 log could be achieved using a practical feed concentration of 1 mg/L. At the same feed concentration and permeate detection limit of 20 ng/L, R-WT challenge levels of >4.5 could be achieved. However, R-WT is not fully rejected by RO and NF membranes, and practical removals of 4 and 3 logs, respectively, for RO and NF membranes can be realized. Given the high prices for microspheres currently, their use at full scale is not practical. Dosing microspheres to a 5 mgd membrane plant to achieve 4-log removal challenge would cost nearly $100,000. In contrast, the same dosing of R-WT would cost less than $500, making R-WT cost effective as a full-scale integrity testing viral surrogate.
- Conductivity and/or TDS rejection cannot be used as an accurate predictor of viral passage. Intrinsic imperfections are not distinguishable using normal manufacturer's QA/QC "wet testing" which simply measures percent rejection of a salt solution.
- The R-WT may be a practical surrogate for detecting imperfections in both RO and NF membranes relative to virus removal. It is inexpensive, easy to dose, and can be accurately measured at concentrations down to 20 ng/L in membrane permeates, resulting in an ability to measure about 4 log removal at a feed concentration of <2 mg/L. R-WT adsorbs in the membrane (to a higher extent with TFCS), requiring that flushing must be performed to remove its presence in the permeate following application. Application of R-WT as a surrogate may not be ideal, however, due to inherent differences in the mechanism by which dye and virus are removed and the limit of rejection by both NF and RO (about 4 logs).
- The R-WT removals decreased somewhat as a function of run time for both element types most likely due to dye uptake within the polyamide rejecting layer and its subsequent diffusion into the permeate. The reduction in R-WT removals with time was more significant with TFCS than ESPA1, which may be explained by the differences in membrane surface chemistry although both are aromatic polyamide membranes. Visual observation showed a greater degree of dye absorption for TFCS.
- Flushing the systems with feedwater for 24 hours was found to be sufficient to reduce R-WT concentrations in the permeate to background levels, which could otherwise cause a positive interference in the measurement of fluorescent microspheres. The amount of flushing time required for full-scale use of R-WT would be a function of the amount of dye permitted in potable water. It is not envisioned that the levels observed in the NF/RO permeates (typically less than 1,000 ng/L) would be a concern.

- In terms of the impact of single element vs. two-stage operation on surrogate removals, the particle surrogates (virus and spheres) were removed slightly greater in MSTU than in DETU, while the opposite was found for R-WT. This may be expected considering the passage of particles through RO or NF membranes is independent of feed concentration. In contrast, for R-WT, the removal is by solution-diffusion resulting in the smaller removals with MSTU due to dependence on feed concentration. This is an important difference between the surrogates and one that must be accounted for when using R-WT in full-scale systems.
- The presence of low concentrations of the tested surrogates during dosing did not cause membrane fouling or modification to the conductivity/TDS rejection characteristics either through cake layer formation (*phage* or microspheres) or adsorption within the membranes (R-WT). This indicates that use of such surrogates at full-scale would not be constrained by fouling considerations.

IMPORTANCE OF SYSTEM INTEGRITY

- The findings associated with the membrane and o-ring compromization highlight the importance of complete integrity of membranes and associated equipment in practical membrane applications in terms of achieving the desired level of treatment. It should be noted, however, that the types of imperfections used in this research have not been correlated with those commonly expected in full-scale plants. It is not uncommon for o-rings to be damaged in full-scale plant operation as elements and interconnectors shift within the vessel during start-up and shutdown. Such shifting has been observed to cause cutting or rolling from the o-rings groove. Imperfections of the size induced in this research are probably uncommon, however the presence of smaller imperfections, at least in the thin film barrier, is certainly possible. If present in sufficient quantity they could potentially produce a similar degree of advective flow.

APPLICATION TO WATER INDUSTRY

The applicability of this research to the water industry can be summarized as follows:

1. ***Rhodamine WT***. Rhodamine WT (R-WT) is currently applicable as a surrogate for evaluating the integrity of full-scale high pressure membrane systems for virus removal. In this regard, R-WT has been included in the American Society for Testing and Materials (ASTM) draft *Standard Practice for Integrity Testing of Water Filtration Membrane Systems* (ASTM 2002). It must be noted that the capability of R-WT to assess viral removal will differ for RO than NF and could differ between manufacturer's products for each process. It is reasonable to assume that 4-log virus removal can be successfully demonstrated with a RO system, but only 3-log removal with a NF system. The reader is referred to the appropriate section of this draft practice for guidelines on the use of R-WT. The actual application of R-WT at full-scale will have to address a number of issues that were identified but not resolved during this testing. These include:

- Approval by relevant State and Federal regulatory agencies regarding R-WT application to a full-scale membrane facility. To date, R-WT has only been approved for use for tracer studies in water treatment plants.

- Time required to flush the system following R-WT application in order to reduce concentrations of R-WT in the membrane permeate to less than 100 ng/L (or other level imposed by regulatory agency). The flushing duration will be a function of log removal by the full-scale membrane system, most notably whether the system employs RO or NF membranes. It is anticipated that flushing time with RO membranes may be very short given that log removals by intact RO membranes as observed in this study produced permeate R-WT well below 100 ng/L. For systems with NF membranes, flushing time will have to be determined based on the NF product used and its actual R-WT rejection.

- Ability to discharge membrane concentrate containing R-WT at concentrations up to 7 mg/L. Assuming a target R-WT dose in the membrane feedwater of 1 mg/L and a system capability of 3-4 logs R-WT removal, the concentrate from the full-scale plant will contain R-WT concentrations of 6-7 mg/L. At this concentration, the dye color may be observable. Depending upon the source of concentrate discharge, the presence of R-WT at these levels may represent a challenge, particularly for surface water discharges. The applicable regulatory agency(ies) should be contacted to determine the implications of R-WT use. It must be kept in mind that the time period for R-WT dosing can be very short (<15 minutes) and as such the mass of microspheres to be discharged will not be significant.

2. ***Fluorescent Microspheres.*** Currently, fluorescent microspheres of the type used in this research are not practical for use by utilities in assessing the integrity of full-scale membrane systems. Although the microspheres showed the best correlation to MS-2 phage with respect to both intact and compromised membranes and systems, their high cost makes their dosing of full-scale facilities prohibitive. Furthermore, current analytical methods needed for their detection require expensive and sophisticated bench-top instrumentation and well-trained technicians. These constraints reflect the development of such microspheres for the biomedical industry, where a high level of precision and quality control/quality assurance in manufacturing is needed. In order to make microspheres applicable to membrane integrity testing at full-scale, it will be necessary to engineer a product that has much lower production costs and is designed specifically for such a use. This will require collaboration between the water industry and either current microsphere manufacturers or other commercial entity with the capability and interest to produce such a product.

3. ***Types of Compromization.*** The types of compromization used in this research are not necessarily those that most utilities would be expected to encounter with full-scale plants. The quality control practiced by most membrane suppliers would make "pin pricks" rare, however, it is certainly possible that membrane imperfections caused by faulty barrier layer coating and polymerization or glue line discontinuities can produce similar advective flow. For o-rings, more common modes of failure relate to partial cuts and cracks from improper installation and shimming (where o-rings can be rolled from the

groove). In this case, the results obtained in this study for the cracked o-rings are most pertinent and indicate that major loss of compromization from o-ring failures should be low.

4. ***Membrane Fouling.*** A major and important finding from this research that has direct applicability to the industry is the ability of membrane fouling to "heal" membrane imperfections and further that such healing could be reversed by chemical cleaning. It was surprising to find that highly pretreated membrane feedwater (feed filtered by a 0.2 um microfilter), where only submicron particles should be present, was capable of bridging a defect on the order of 300-500 microns. Given that most membrane imperfections induced during the membrane/element manufacturing process should be less than this size, it is reasonable to assume that viral passage through imperfections in elements received from the manufacturer should experience decreased viral passage with increased operation. This has been observed for many full-scale plants with respect to conductivity/TDS passage and is termed "membrane conditioning" in the industry. Such operational improvement in virus removal is advantageous and with sufficient sampling and statistical analysis, could provide the basis for ensuring that full-scale systems receive appropriate virus removal credit following some initial conditioning period. This, of course, assumes that the full-scale system contains sufficient imperfections as not to meet regulatory requirements upon start-up.

APPENDIX A

Appendix A contains figures or tables that are not presented in Chapter 2 and are related to materials, methods and results of bench-scale testing portion of this project.

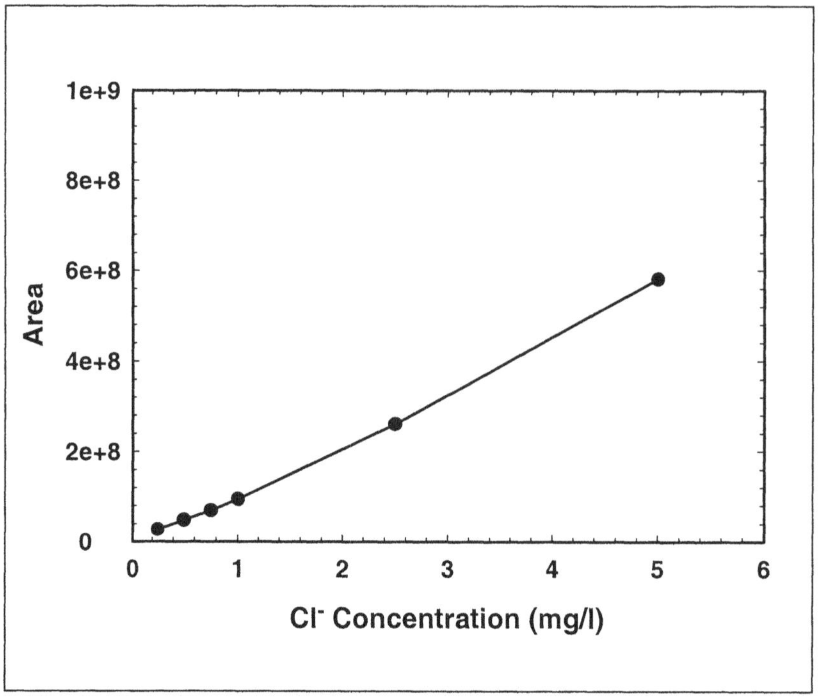

Figure A.1 Typical ion chromatography calibration curve for chloride ion

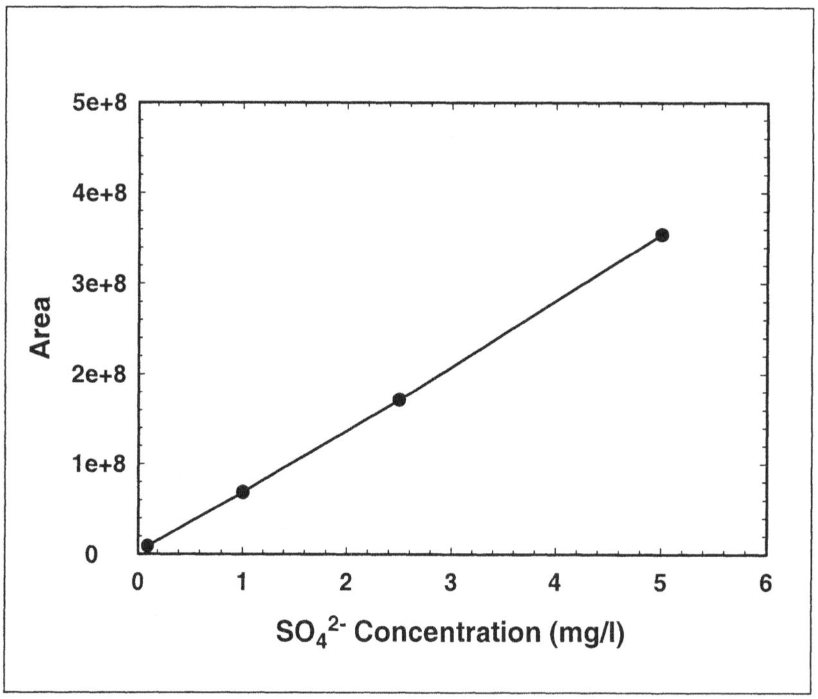

Figure A.2 Typical ion chromatography calibration curve for sulfate ion

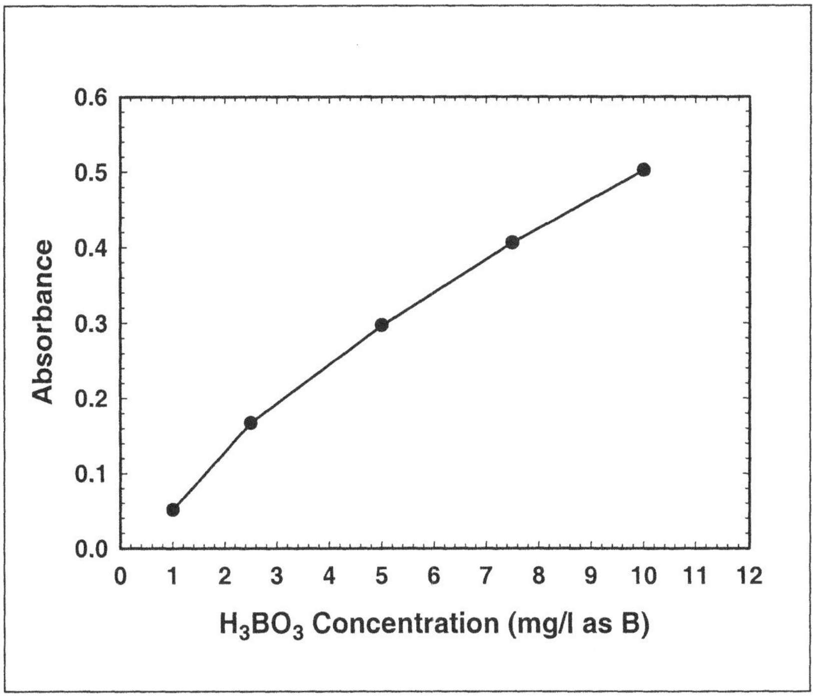

Figure A.3 Typical spectrophotometry calibration curve for boric acid (Carmine Method)

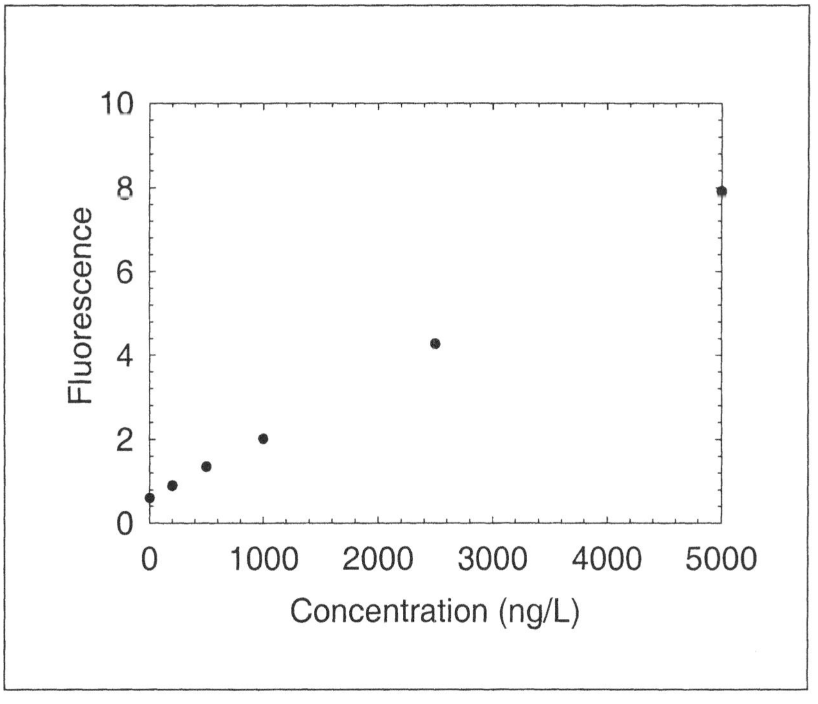

Figure A.4 Representative calibration curve used for the determination of microsphere concentrations in permeate samples from flat-leaf and spiral-wound element experiments

APPENDIX B

Appendix B contains figures or tables that are not presented in Chapter 3 and are related to materials, methods and results of pilot-scale testing portion of this project.

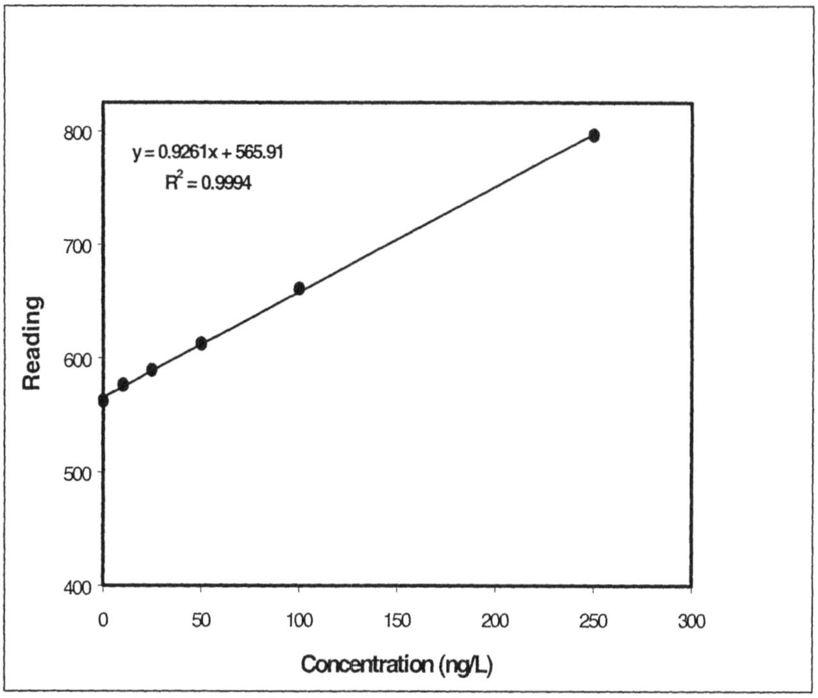

Figure B.1 Representative R-WT standard calibration curve for ng/L range

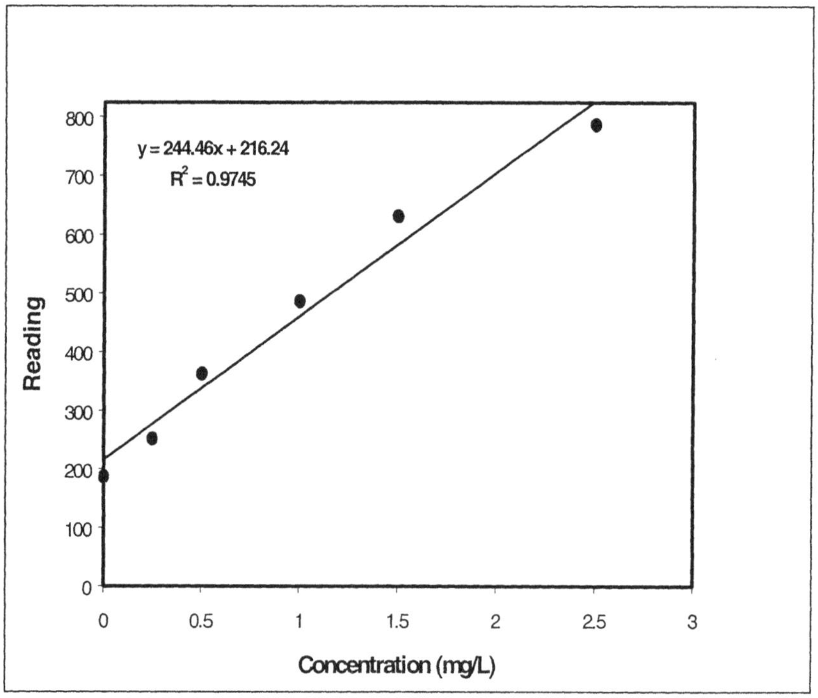

Figure B.2 Representative R-WT standard calibration curve for mg/L range

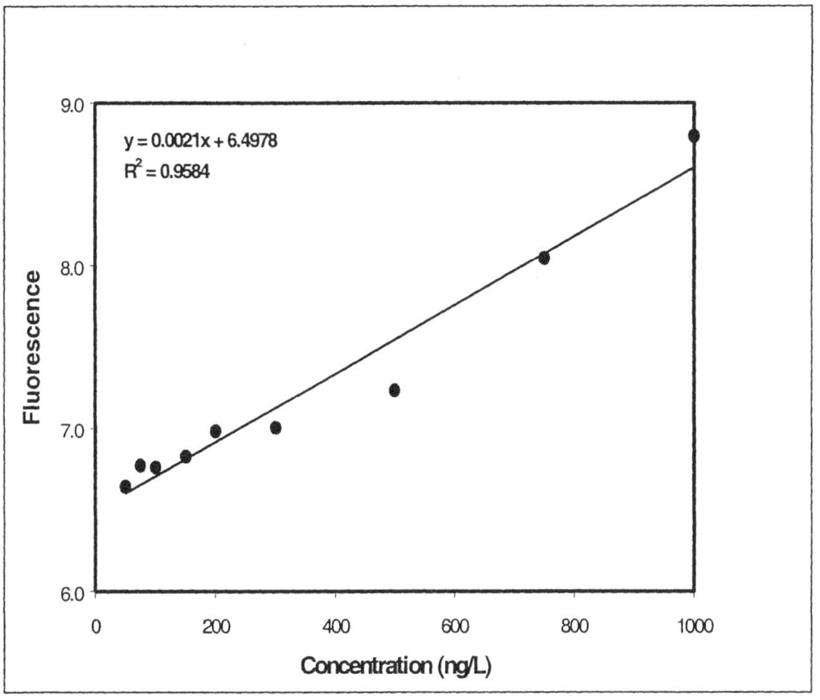

Figure B.3 Representative microsphere standard calibration curve for ng/L range

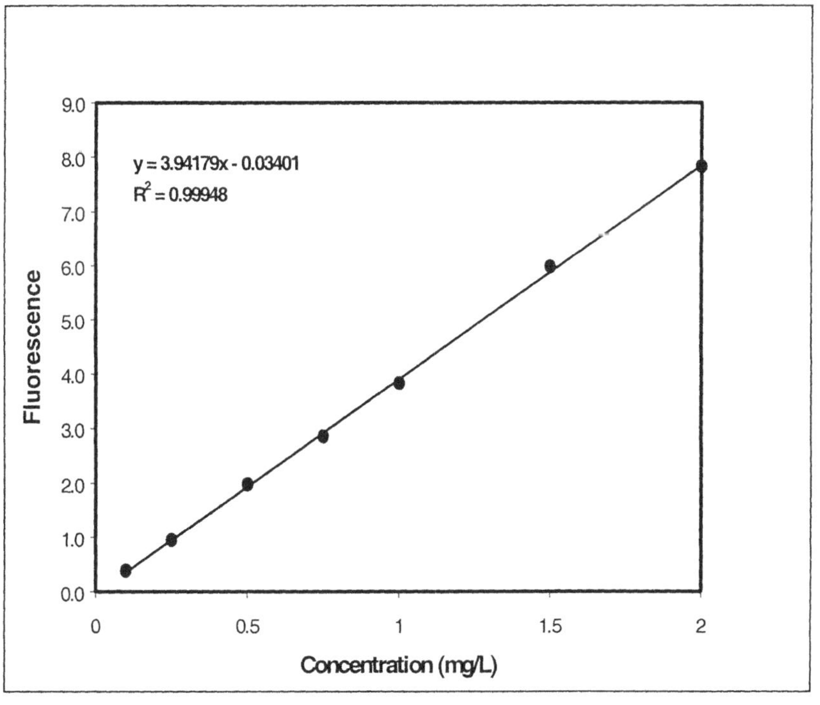

Figure B.4 Representative microsphere standard calibration curve for mg/L range

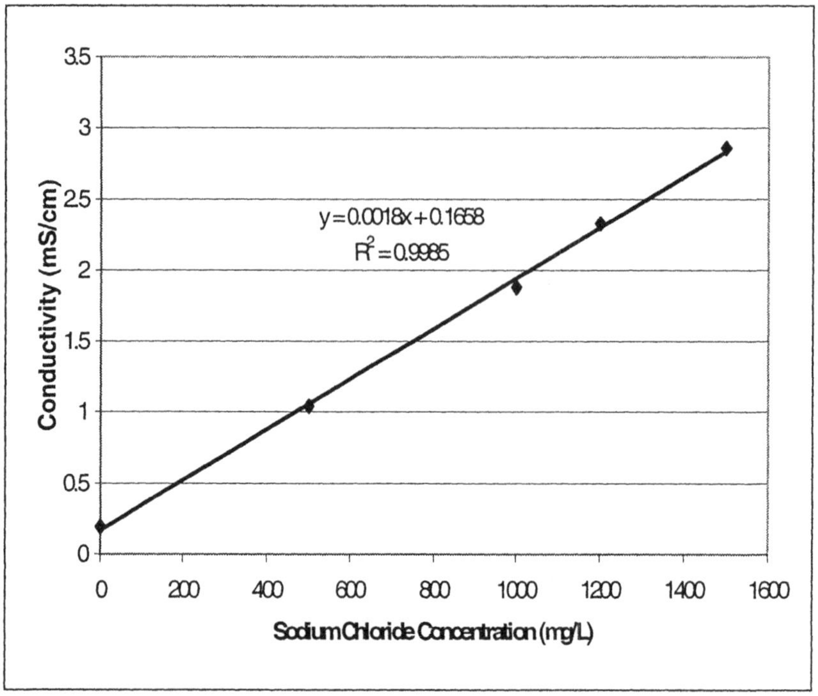

Figure B.5 Representative sodium chloride standard calibration curve

REFERENCES

Adham, S.S., R.S. Trussell, P.F. Gagliardo, and R.R. Trussell. 1997. Virus Removal by Reverse Osmosis Membranes. In *Proc. of Annual AWWA Conference*. Atlanta, GA: AWWA.

Adham, S.S., R.S. Trussell, P.F. Gagliardo, and R.R. Trussell. 1998a. Rejection of MS2 Virus by RO Membranes. *Jour. AWWA*, 90(9):130-135.

Adham, S.S., P.F. Gagliardo, D. Smith, D. Ross, K. Gramith, and R. Trussell. 1998b. Monitoring of Reverse Osmosis for Virus Rejection. In *Proc. of the AWWA Water Quality Technology Conference*. San Diego, CA: AWWA.

APHA, AWWA, and WEF (American Public Health Association, American Water Works Association, and Water Environment Federation). 1992. *Standard Methods for the Examination of Water and Wastewater*. 18th ed. Washington, D.C.: APHA.

ASTM (American Society for Testing and Materials). 1988. *Standard Practices for Detecting Leaks in Reverse-Osmosis Devices*, Section 11, D3923-80. Philadelphia, PA.: ASTM.

ASTM (American Society for Testing and Materials). 2002. *Standard Practice for Integrity Testing of Water Filtration Membrane Systems*. ASTM Task GroupD19.08.02. Revision J.

Brian, P.L.T. 1966. Mass Transfer in Reverse Osmosis (Chapter 5). In *Desalination by Reverse Osmosis*. Cambridge, MA: MIT Press.

Chiou, C-F, M. Torres-Lugo, B.J. Marinas, and J.Q. Adams. 1997. Non-Biological Surrogate Indicator for Assessment of Ozone Disinfection Efficiency. *Jour. AWWA*, 89(8):54.

Filmtec Corporation. 1997. *Product Information for FILMTEC Membranes*, 8-inch BW30-330 Brackish Water RO Element. Midland, MI.

Fluid Systems. 1995. *Product Bulletin for TFCR Spiral-Wound Reverse Osmosis Element*, Brackish Model TFCR 8822HR High Rejection Polyamide Membrane.

Gagliardo, P.F., S.S. Adham, and R.R. Trussell. 1997a. Water Repurfication Using Reverse Osmosis: Thin Film Composite vs. Cellulose Acetate Membranes. In *Proc. of the AWWA Membrane Technical Conference*. New Orleans, LA: AWWA.

Gagliardo, P.F., S.S. Adham, R.R. Trussell, and A.W. Olivieri. 1997b. Membranes as an Alternative to Disinfection. In *Proc. of Annual AWWA Conference*. Atlanta, GA: AWWA.

Jacangelo, J.G., S.S. Adham, and J.M. Laine. 1995. Mechanism of Cryptosporidium, Giardia and MS2 Virus Removal by MF and UF. *Jour. AWWA*, 87(9):107-121.

Lonsdale, H.K., U. Merten, and R.L. Riley. 1965. Transport Properties of Cellulose Acetate Osmotic Membranes. *Journal of Applied Polymer Science*, 9:1342.

Millipore Corporation. 1998. Personal Communication.

Molecular Probes. 2000. *Product Information*. Eugene, OR.

Moll, D. J. 2002. Dow Chemical Company. Personal communication with Jim Lozier.

Orange County Water District (OCWD) and James M. Montgomery, Consulting Engineers, Inc. 1979. *Water Factory 21 Virus Study*. Fountain Valley, CA: OCWD.

Rose, J.B. and S.R. Farrah. 1993. Microbial Monitoring, Pathogen Removal Assessment and Public Health Risks Associated with Reclaimed Water. Advanced Water Reclamation Program Final Report Vol. II for Reedy Creek Improvement District, Lake Buena Vista, FL. CH2M HILL, Inc., Denver, CO.

Sorber, C.A., J.F. Malina, and B.P. Sagik. 1972. Virus Rejection by the Reverse Osmosis Ultrafiltration Processes. *Water Research*, 6:1377-1388.

Spangenberg, C. 1997. Irvine Ranch Water District, CA. Personal Communication with Jim Lozier.

State of California Dept. of Health Services. 2000. *California Code of Regulations, Title 22.*

Trussell, R.R., T.G. Richardson, S.S. Adham, and I.N. Najm. 1995. Water Repurification Project- Phase 2 Advanced Treatment Pilot Study, City of San Diego, CA. Montgomery Watson, Pasadena, CA.

Turner Designs. 2000. *Bulletin 104*. Sunnyvale, CA.

Urama, R.I. and B.J. Mariñas. 1997. Mechanistic Interpretation of Solute Permeation Through a Fully Aromatic Polyamide Reverse Osmosis Membrane. *Journal of Membrane Science,* 123:267-280.

U.S. Environmental Protection Agency. 1989. The Surface Water Treatment Rule. *Fed. Reg.* 54FT27485.

U.S. Environmental Protection Agency. 1990. *Guidance Manual for Compliance with the Filtration and Disinfection Requirements for Public Water Systems Using Surface Water Sources.* Contract No. 68-01-6989.

U.S. Environmental Protection Agency. 1997. *Guidance Manual for the Surface Water Treatment Rule.*

Winograd, Y., A. Solan, and M. Toren. 1973. Mass Transfer in Narrow Channels in the Presence of Turbulence Promoters. *Desalination, 13*:171.

Yahya, M.T., C.B. Cluff, and C.P. Gerba. 1993. Virus Removal by Slow Sand Filtration and Nanofiltration. *Water Science and Technology*, 27(2-4):445.

ABBREVIATIONS

ASTM	American Society for Testing and Materials
AwwaRF	Awwa Research Foundation
CA	cellulose acetate
COMP	interfacially formed composite membrane
CTRW	conventionally treated river water at NWR-WTP
CV	coefficient of variation
D/DBP	Disinfectants/Disinfection Byproduct Rule
DDW	distilled and deionized water
DETU	dual-element test unit
EPA	United States Environmental Protection Agency
ESWTR	Enhanced Surface Water Treatment Rule
gfd	gallon per square foot-day
gph	gallon per hour
gpm	gallon per minute
HP	high-pressure
IRWD	Irvine Ranch Water District
log	logarithmic
MF	microfiltration
MFSE	microfiltered secondary effluent at WF-21
mgd	million gallon per day
MSTU	multi-stage test unit
MVTU	multi-vessel test unit
MWCO	molecular weight cut-off
NDF	net driving force
NF	nanofiltration
NTU	nephelometric turbidity unit
NWR-WTP	Northwest River Water Treatment Plant
OCWD	Orange County Water District
pfu	plaque forming unit
ppb	parts per billion
ppt	parts per trillion
psi	pounds per square inch

RO	reverse osmosis
rpm	revolution per minute
R-WT	Rhodamine WT dye
SBS	sodium bisulfite
SETU	single-element test unit
SWTR	Surface Water Treatment Rule
QA/QC	Quality Assurance and Quality Control
TDS	total dissolved solids
TOC	total organic carbon
TYE	tryptone yeast extract
UF	ultrafiltration
UVA	ultraviolet light absorbance
WF-21	Water Factory-21

LIST OF SYMBOLS

A	water transport coefficient
B	solute transport coefficient
c	solute concentration
D	solute molecular diffusivity in water
J_s	gravimetric solute flux
J_v	volumetric solution flux
k	solute mass transfer coefficient
p	hydraulic pressure
P_f	system feed pressure, psig
P_p	permeate side pressure, psig
μm	micron
μS	microsiemens
ε	fraction of total permeate passing the membrane through imperfections
π	osmotic pressure of the solution
δ	thickness of the concentration polarization layer
ΔP_{fc}	difference in pressure between feed side of first stage and concentrate side of last stage, psi
$\Delta \Pi$	difference in osmotic pressure across the membrane or between feed/concentrate and permeate, psi

IWA Publishing's authorised EU representative for General Product Safety Regulations is Diane D'Arras, 15 rue Duret, 75116 Paris, France, e-mail: safety@iwap.co.uk.

Printed and bound by CPI Group (UK) Ltd, Croydon, CR0 4YY
17/09/2026
02218793-0005